张鸿室内设计作品集 — A Collection of Interior Design Works by ZhangHong

Feel Freedom Forever

张鸿 著

华中科技大学出版社

图书在版编目（CIP）数据

味道长：张鸿室内设计作品集 / 张鸿 著. —武汉：华中科技大学出版社，2011.7
ISBN 978-7-5609-6871-1

Ⅰ.①味… Ⅱ.①张… Ⅲ.①室内设计 - 作品集 - 中国 - 现代 Ⅳ.①TU238

中国版本图书馆CIP数据核字（2011）第004850号

味道长：张鸿室内设计作品集　　　　　　　　　　　张　鸿 著

出版发行：华中科技大学出版社
地　　址：武汉市武昌珞喻路1037号（邮编:430074）
出 版 人：阮海洪

策划编辑：成海沛　　　　　　　　　　　　责任监印：张贵君
责任编辑：吴红敏　　　　　　　　　　　　书籍设计：王　江

印　　刷：北京利丰雅高长城印刷有限公司
开　　本：965mm×1270mm 1/16
印　　张：18.25
字　　数：150千字
版　　次：2011年7月第1版 第1次印刷
定　　价：238.00元

投稿热线：(010)64155588-8000 hzjztg@163.com
本书若有印装质量问题，请向出版社营销中心调换
全国免费服务热线：400-6679-118 竭诚为您服务
版权所有 侵权必究

张
鸿 Zhang Hong

中国建筑装饰协会设计委员会副主任委员
中国建筑设计研究院张鸿酒店设计机构负责人
英国H.H.D国际酒店设计顾问公司负责人

China Building Decoration Association Deputy Chairman of the Design.
Honghouse Head of Building Decoration Engineering Co., Ltd.
The person in charge of H.H.D International Hotel Design Consulting Corporation, Ltd. of England

序言
Preface

味道长

"味道长",是源于巴蜀民间美食家品评绝妙菜品时的一句意犹未尽的口头禅,具惬意悠然之妙义,无法以言语形容。张鸿先生以此为号,并将空间设计作品专集以此为名,却另有深意。

"味",知味、体味、意味、人生百味……味是一种气韵,一种体验,难以文字表述。"道",在中国古典哲学范畴中,是蕴涵于天地万物之中,无间不入,无所不包,却又视之不见、听之不闻、搏之不得的一种存在,是万物的本源,是万物发展变化的规律。在现实中,无法描述,只能用心灵感知。

"长",一为久远,一为深长。由过去,到现在,至未来,无息无止。味道长,是他对人生的感悟,是对空间设计的深刻诠释,是毕生孜孜追求的境界。

知 味

千年的巴蜀古镇,清溪秀丽,建筑古朴,庭院深深,从懵懂少年开始就吸收着古镇悠然自得的文脉和尔雅气息。一生难以忘怀的是院中黄角树下慈母的笑容;奔腾的雅砻江带着青春的热情与豪迈,领略着川西南山脉的挺拔与灵秀,以及大地母亲的包容与和谐,显得生机勃勃;嘉陵江上化不开的氤氲云雾,夹杂着老火锅的醇香,恍如混沌初开之际,在物我交融中思索生命的本源,体悟"存在"与"不存在","空"与"非空"……

大自然的天然熏陶,塑造了他"物我交融"的情感和"天人合一"的性格。大自然的百态百味,也开启了他的"知味"之旅。

体 味

体味人间却远没有这般惬意。从油画创作,到空间设计,到工程实施,再到空间设计。做学问,教学生,高低浮沉,曲折变迁。不变的,是一颗求索的心。

"体味"的过程是一条塑造智慧的路,他尝试让一种神圣的东西得以体现。现实的矛盾没有淹没他,却一直成为他进取的源泉。坚韧不拔又要顺其自然,中庸与多重性,在山水间他不是一个随波逐流的过客,而是满怀虔诚的苦行僧。

意味悟道

"体味"的过程异常艰辛,令人刻骨铭心。是重负也是厚积,在"意味"深长中渐悟。逐渐地,他体悟了生命之真义和性灵之超越,如"气"般融入自然,不受任何实体束缚,与宇宙自然一同运行。这便是"道"。

无论是生存之道,经营之道,品茗之道,烹调之道,还是设计之道。将空间设计生命化,将所有事物融为一体,无形无相,每一种生灵都自在而怡然地存在着,各具神韵又和谐而美好,生命美学即为自然美学。

在体味空间的那一刻,生命的悦动,令灵感自然升腾。创作不再是刻意的,而是自然的。无意识创作,从容坦荡,清朗亮丽而透明,一种酣畅淋漓的境界油然而生。因百"味"而悟"道","味"因"道"而生动。

修 道

空间设计有别于艺术创作，不是表现自我。看似洋洋洒洒随心所欲，其实法度严谨。空间是"器"，不是简单粗暴的容纳色彩、材料以及各类物件，而是"生命之器"，是将有缘之性灵聚合，包括空间的主人，和谐依赖，和乐相濡。或古朴或现代、或简约或繁复，无所谓中西古今，和合众缘，只需气息相融。虽气味不同，却因道而投合。"有我"而"无我"，"有形"而"无形"，浑然一体，气韵生动、通泰顺畅，毫无涩滞混浊之感。

如此，空间设计师便是渡缘之人，也是循道调味之人。透过技巧与素材的使用调度，增益减损的调和，刚柔并济，天地方圆，呈现"外师造化、中得心源"的至高境界，从而使建筑空间与生活、性灵交融，此方为空间设计的真意。

因而，设计师不可将自我意识随意强加，不应有风格，或者不应陷入某一格。在"无为"中"有为"，在"无我"中找到自我。喜爱某味而不贪恋一味，以舟渡河而不过河后负舟前行，"善用"而不"苦执"。正如苏东坡所言"行于所当行，止于所不可不止"，这便是设计师的修行。

正本清源，修身以循"道"。

传 道

空间设计师的使命，不仅仅是设计。

当今人类受物质文明的冲击而陷入精神苦闷，性灵迷乱而飘零，以何为归依？以何为引领？如何觉悟？如何恢复本性生智慧？千手千眼观世音渡众生，因人施教，没有定法，因而有无边妙法，令众生修"定"后开"慧"。设计师的每一次空间设计，即是在立一个方便法门。

或繁或简，解"味"循"性"而入，以个人情感为归依构建和谐舒适、令人融入共鸣的空间，心境得以平和安宁、意念得以凝练，接近禅的沉思境界。若更能以妙法引磬入境，体味雨露清泉自然百味，加深人与自然相合相契的可能，便有机缘引领居者回归万物合一的原初自然，由迷失走向觉悟，恢复本能，得以福慧双修。

超越独善而我执的"小爱"，从自然的灵魂中寻找力量，在弘扬沉雄博大圆熟的中华传统文化中获得灵慧，以更大的魄力去矫正时弊，透过设计这一载体发扬"大乘"精神，这便是空间设计师的使命。

以"身"载"道"，以"味"传"道"。

回 归

粹然而真，淡然而纯，浩然而正，至善的天然性情，唯有赤子之心，方得回归自然。以一种无我、包容、宽宏的气度，在旷达深奥的天地间，悠然展翼，飞扬自在，以智慧迎向未来，开创无限生机。鸿，翔于天，归于地，灵真之气，永生不灭。

一个好作品，可以构建一个感应自然之器；
一个好设计师，可以在修行中阐扬文化，开启回归自然之契机；
一本好书，可以成就一个性灵，遇到知己，欣然契合。
品味，悟道，天地永恒。

张亚莉　Zhang Yali

Feel Freedom Forever

The "Feel Freedom Forever" is born of the pet phrase of tasty summary that is often used by the Bashu (Sichuan) folk gastronomes in their commenting on delicious dishes, which has a wonderful meaning of satisfaction and leisureliness that cannot be fully expressed by words. It is, however, of another profound meaning for Mr. Zhang Hong to take it as his assumed name and to use it as the title of collection of his space design works.

The "taste" involves in taste of realization, taste of experience, taste of consciousness and different tastes of life… The taste is an artistic conception and an experience, which could hardly be described by words. In the classical Chinese philosophical category, the "Dao" is an existence contained in the universe where there is no space to be inaccessible and nothing to be exclusive but could hardly be seen, heard and obtained. And it is the origin of all things on earth and the law of development and changes of all the things. But in reality, it could only be perceived by soul but could hardly be described. The "lasting" means both the far-back and the far-reaching. From the past, until now and to the future, it goes with no stop and no end.

The "Feel Freedom Forever" is his perception for human life, his profound interpretation of space design and his realm of lifelong persevering pursuit.

Taste of Realization

In an ancient Bashu (Sichuan) town with a history of more than one thousand years, and with beautiful lucid rivers, unsophisticated architecture and deep courtyards, he began from his ignorant early youth absorbing the leisurely cultural tradition and cultivated breath of the ancient town, and he could not forget in his life the smile of his loving mother under the yellow horn tree in the courtyard. The impetuous Yalong River flows with youthful enthusiasm and heroism, which appreciates the magnificence and smartness of the mountains in the southwest Sichuan and tolerance and harmony of the Earth Mother, and everything there is full of vitality. The dense mist above the Jialing River, mingled with the mellow of old chafing dish, is something like the occasion of opening the chaos. In the communion between human feelings and secular world, he has speculated on the origin of life and appreciated the "existence" and "non-existence", "empty" and "non-empty"…

The natural influence from Nature has shaped his emotions of "communion between human feelings and secular world" and character of "unity of heaven and humanity". The different postures and tastes of Nature make him start his tour of "taste of realization".

Taste of Experience

The experience of human world is far from such a satisfaction. From creation of oil paintings to space design, to project implementation and then back to space design, from scholarship engagement to students teaching, he has experienced various ups and downs and many winding changes. What keeps constant is his heart to pursue.

The process of "Taste of Experience" is a route to shape wisdom, and he has tried to make a sacred thing be realized. The realistic contradiction does not submerge him, but has always become the source of his impulse. With perseverance and natural tendency, and with moderation and multiplex, he is not a passing traveler who goes adrift in the landscape, but an ascetic monk with full piety.

Taste of Consciousness and Perception of Dao

The process of "Taste of Experience" is full of abnormal hardships, which is unforgettable. Either a heavy burden or a thick accumulation would be gradually perceived in the profound "Taste of Consciousness". Gradually, he has understood the real meaning of life and surpassing of spirit, which is integrated with nature as the "air" free of any entity bondage and does operate together with the universe and nature. This is the "Dao".

It goes the same no matter it is the Dao of existence, the of business, the Dao of tea drinking, the doctrine of cooking or the Dao of design. To make space design more life-emphasized and to blend everything into one with intangibility and formless, each kind of spirits exists in a free and happy way, and they have their own romantic charms but harmonize with great happiness. Life aesthetics is natural aesthetics.

At that moment of space experience, the pleasant move of life makes the inspiration naturally rise. Then creation is no longer painstakingly, but would be natural. The unconscious creation is calm, magnanimous, clear, bright and transparent, which produces a realm with ease and verve. The "Dao" is perceived because of different "tastes", and the "tastes" would be vivid because of the "Dao".

Cultivation of Dao

Space design is different from artistic creation, and it is not a self-expression. It seems at great length voluminous and at one's own will, but in fact it has a strict law. Space is a "container", which is not simple and crude accommodation of colors, materials and various objects, but it is the "container of life". It would polymerize all the destined spirits, inclusive the owner of the space, to depend upon each other with great harmony and happiness. No matter it is simple or contemporary, concise or complicated, Chinese style or Western style, ancient or modern, the synergy of conditions needs only to breathe in harmony. Although their smells are different from each other, they could indeed get along because of the Dao. From "selfness" to "selflessness" and from "tangibility" to "intangibility", all of them blend into one harmonious whole, with vivid artistic conception and unobstructed smoothness, which has no sense at all of acerbity and dimness.

Therefore, a space designer is the man to save destiny and also the man to follow the doctrine and make the taste. Through the use and control of technique and materials, the mediation of gain and loss, the combination of hardness with softness, the consideration of the universe and circumstances, he would present the supreme realm "to learn outward from nature and to get inward from heart", so as to make the architectural space blend with life and spirits, which would be the intendment of space design.

Thus, the designer could not impose, at will, his self-awareness, and should not have his own style or fall into one particular style. There should be "activity" in "inactivity" and "selfness" from "selflessness". He could love one taste but would not cling to it, and he could cross the river by a boat but would not move forward by carrying the boat. He would "make a perfect use" but not "make a suffering clinging to it".Just as Su Tung-po said, "To do what you should do and stop what you could not continue". This is the cultivation of a designer. To make a radical reform, he should cultivate his moral character to follow the "Dao".

Propagation of Dao

The mission of a space designer is not simply the design.

As the mankind falls into spiritual torture because of the impact of material civilization today, the spirits are confused and homeless. What is the refuge? What is the lead? How to be enlightened? How to restore the natural instincts to produce wisdom? The Guan-yin Bodhisattva with One Thousand Hands and Eyes saves all living creatures with different treatment in salvation but no fixed law. Therefore, there is the boundless wonderful Dharma to order all living creatures to make "cultivation" before they open "intelligence". Each of the space designs of designers would make a convenient Pharman.

No matter it is simple or complicated, the solution of "taste" should be done according to the "nature". To construct, with personal emotions as the refuge, a space of harmonious and comfortable that could make people blend in resonance, the mood could be made peaceful and tranquil and the idea could be refined to reach close to the Zen realm of meditation. If the wonderful Dharma is further used to lead the inverted bell to enter to experience such natural tastes of different kinds as rain and dew and clear spring, and promote the possibility for human and nature to coincide and unite, there would be a lucky chance to lead the habitants to return to the original nature of unity of all things on earth, thus they could go from disorientation to consciousness and restore their instincts so as to make a dual cultivation of both happiness and intelligence.

It is the mission of space designers to surpass the "little love" of virtuous self conduct and egocentrism, to seek strength from the soul of nature, to gain the wisdom from the Chinese traditional culture that is of calm and magnificent development and profound and proficient, to rectify the current malpractice with even greater boldness and to promote the spirit of "Great Vehicle" through the carrier of designs, and this is the mission of a space designer.

It is to carry "Dao" with "mind" and to propagate "Dao" with "taste".

Return

Purity makes genuine, insignificance makes simplicity and spaciousness makes integrity. The holy natural disposition could only be made return to nature by the human natural kindness. With a tolerance of selflessness, forgiveness and munificence, in the broad and profound world, he leisurely spreads his wings and flies at ease to march forward with wisdom to the future and create the infinite vitality. The swan goose flies in the sky and returns to the land, and his spirit of wisdom and purity would exist forever.

A good piece of work could be used to construct a container to interact with nature;

A good designer could propagate culture in his cultivation and open up the opportunity to return to nature; and

A good book could perfect a spirit in his success, who would pleasantly get communion with the bosom friends once he meets.

To savor the taste and perceive the doctrine, you would find the eternal heaven and earth.

目录
Contents

- 004 / 序言　Preface
- 010 / 兜率天宫　Tusita Heaven Palace
- 024 / 家富半岛国际大酒店　Jiafu Peninsula International Hotel
- 054 / 大观园酒店　Grand View Garden Hotel
- 086 / 6号院酒店　No.6 Courtyard Hotel
- 100 / 德州国宾馆　Dezhou State Guesthouse
- 112 / 天和国际大酒店　Tianhe International Grand Hotel
- 130 / 首钢酒店　Shougang Hotel
- 136 / 燕都酒店　Yandu Hotel
- 142 / 某会所酒店　A Club-type Hotel

170	**懿赐会所**	Yici Club
184	**9号会见厅**	No.9 Reception Hall
190	**199天宫会所**	No.199 Heaven Palace Club
208	**梦会所**	Dream Club
220	**萨尔斯堡别墅**	Salzburg Villa
230	**鸟巢文化主题餐厅**	The Bird's Nest Cultural Theme Restaurant
244	**蓝湖郡别墅**	Lanhujun Villa
266	**神华集团神包矿业行政办公楼及联合建筑** Office Building and United Building of Shenhua Baotou Mining Group	
278	**美丽之冠七星级酒店售楼中心** Sales Center of Beauty Crown Seven-star Hotel	
286	**后记**	Postscript

兜率天宫
Tushita
Heaven Palace

项目名称_ **兜率天宫**	Project Name_ **Tushita Heaven Palace**
项目类别_ **宗教建筑**	Project Category_ **Religious Building**
设计面积_ **10 000 m²**	Design Area_ **10,000 m²**
项目地点_ **浙江绍兴**	Project Location_ **Shaoxing, Zhejiang Province**

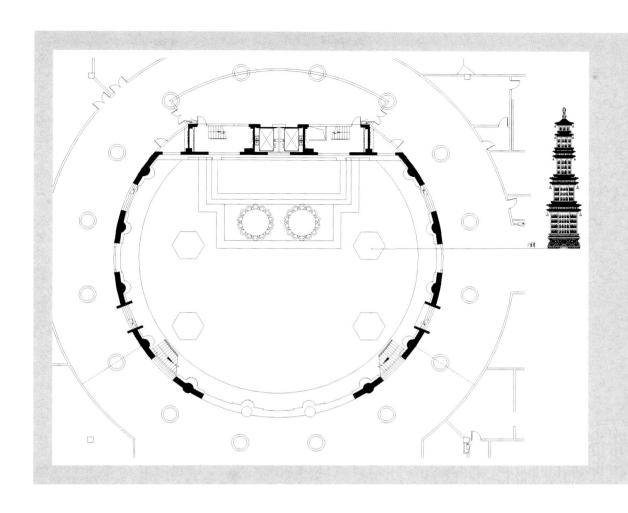

位于浙江绍兴境内的会稽山，是中国著名的历史文化名山。古往今来，这片真山真水孕育和滋养了无数品格高洁之士，积淀了深厚绵长的中国文化。千年桂香茂林修竹，寂静而清新；龙华寺古刹暮鼓晨钟，袅袅香火与云雾齐蒸腾；王羲之兰亭一会，文心与山水共千秋。自然之美、圣坛之灵、文化之韵在此交融，形成了天、地、人和谐共生的清灵大气之境。在此兴建弥勒净土，天人造化，和合众缘。

在佛学中，由娑婆凡尘往生天宫净土是众生修持时现实与未来的超度转化过程。

Located in Shaoxing, Zhejiang Province, the Kuaiji Mountain is one of China's famous historical and cultural mountains. Throughout the ages, this stretch of real mountains and real water has gestated and nourished innumerable people of noble and unsullied character, and it has accumulated the profound and lasting Chinese culture. The millennium cinnamon, flourishing woods and tall bamboos are silent and fresh. In the ancient Longhua Temple with evening drum and morning bell, the curling burning incense curls upwards with the mist; Wang Hsi-chih's visit to the "Orchid Pavilion" demonstrates the common share of the literary mind and the landscape in centuries. The communion here of the natural beauty, the Altar spirit and the cultural charm has produced a territory with a clear and spiritual atmosphere where heaven, earth and human co-exist in harmony. It would be the good fortune of heaven and human and the synergy of conditions to build such a Maitreya's Pure Land right here. In Buddhism, the rebirth from the Saha World to the Tushita Pure Land is the current and future transformational process to release souls from purgatory of all living creatures in their Dharma practice.

本案中巨型的金身佛像、大尺度的空间格局、繁复大气又精雕唯美的须弥宝座，昭示了天宫的殊胜庄严。大体量应用红木纹洞石、实木、金箔，配以从佛学文案中抽象出来的云石拼花地面，整个空间显示了聚集、统领、仰视的空间感悟，并聚合成强烈的感召气场。使众生得以在顶礼膜拜中升腾出天上人间轮回转化的美妙感受，在精神的超度中信步天庭。

高科技全息技术的应用完美呈现了妙丽无比的天宫胜境。在天腾舞的巨龙口喷真气，化为八色水雾，雾中七宝莲花若隐若现，由莲花化生的九亿天子、五百万亿天女翩翩飞舞，众宝曼妙而歌，唱出"苦、空、无常、无我的"梵音法语。声、光、色、形、意……在有形无形，色与空、物与我的转化中，众生灵魂顺缘聚合。

心净娑婆净，天宫亦在人间，人间亦有净土。

In this case, the giant golden Buddha, the large scale of spatial pattern and the Sumeru throne that is complicated, generous, well-carved and aesthetic demonstrate the superior elegance and solemnity of the Heaven Palace. With the huge-scale application of red wood grain travertine, solid wood and gold foil, which is matched with the marble ground with such designs as abstracted from the Buddhist documents, the whole space shows a spatial inspiration of gathering, commanding and upward looking and has polymerized an intense inspiring aura. It makes all living creatures produce, in their worship, the wonderful feeling of samsara and transformation in heaven on earth to stroll to the Court of Heaven in their spiritual release of souls from purgatory.

The application of high-tech holographic technology makes a perfect presentation of the beautiful and superlative scenery of the Heaven Palace. The giant dragon dancing in the sky spurts genuine qi from his mouth, which is then transformed into eight-color water mist. In the mist, there is a looming seven-jeweled lotus, which creates 900 million Sons of Heaven and half quadrillion fairies, who are dancing gracefully in the air and singing joyfully in Buddhist Sanskrit and Dharma language of "duhkha, emptiness, impermanence and selflessness". Sound, light, color, shape, meaning… In the transformation of tangibility and intangibility, expression and emptiness, human feeling and secular world, the spirits of all living creatures come to polymerize in favorable conditions.

Clear mind makes clear Saha World. The Heaven Palace is also in the Mortal World, where the Pure Land does exist.

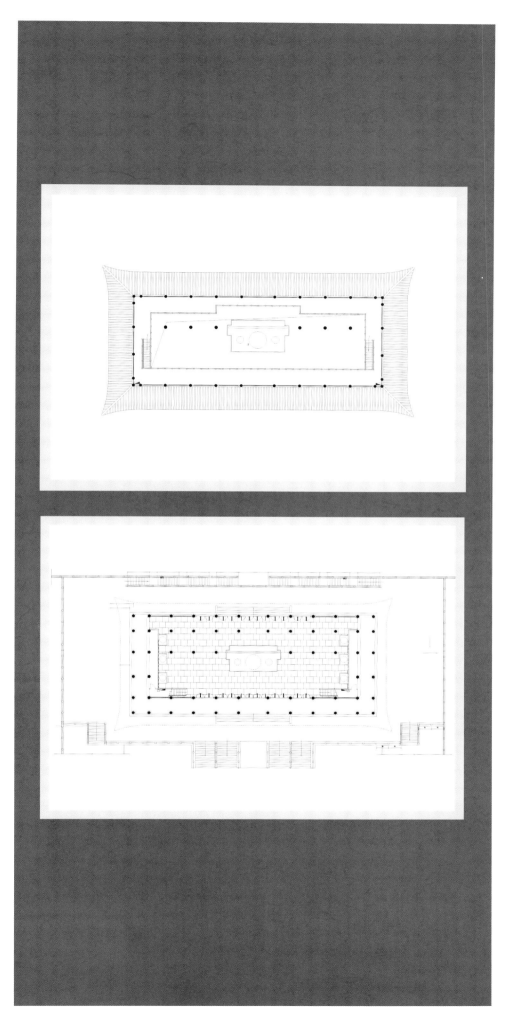

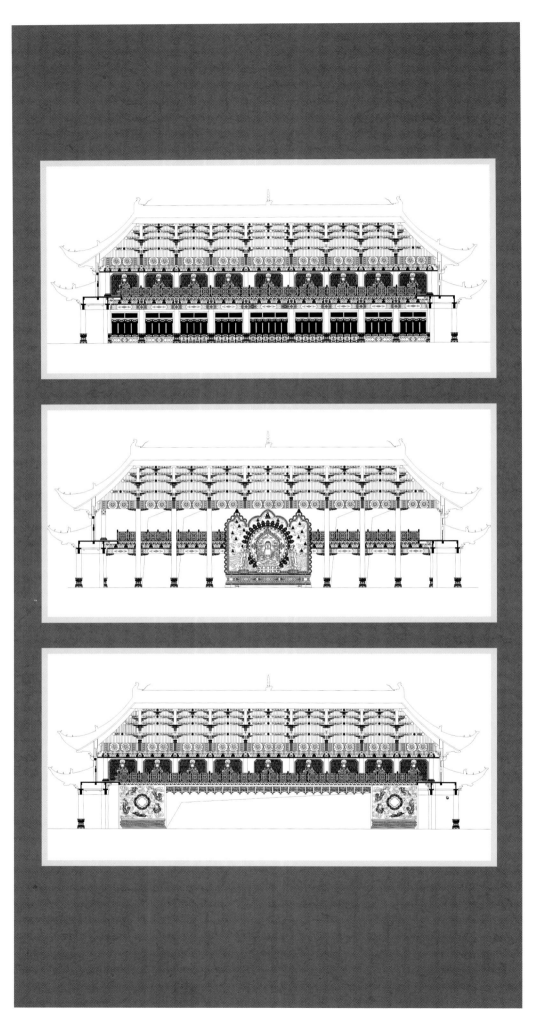

家富半岛国际大酒店
Jiafu Peninsula International Hotel

项目名称 _ 家富半岛国际大酒店	Project Name_ Jiafu Peninsula International Hotel
项目类别 _ 酒店	Project Category_ Hotel
设计面积 _ 100 000 m²	Design Area_ 100,000 m²
项目地点 _ 重庆	Project Location_ Chongqing

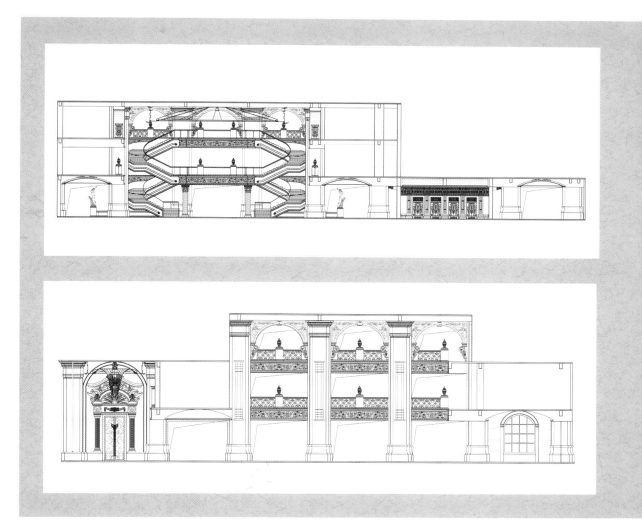

富是基础，贵是升华。贵，不单单只包含物质满足，更包括了文化、品味、素质，因而贵更有内涵。"由富而贵，直至尊贵"是我们设计家富国际大酒店的最基本理念。这也是当今中国财富阶层最强烈的心理期盼。尊贵的气质来源于内外兼修。其外在感观，不仅是金碧辉煌，更应有诱人的气势。酒店的公共空间，以最纯的欧式经典呈现出梦幻般的奢华，但这种奢华并不是繁杂的堆砌，而是以干净流畅的神韵统领着各种豪华的元素，使整个空间具有典雅华贵的气质，更拥有皇宫般居高临下的权威感。伫立于此，一掷千金的豪情，刻意隐富的低调，都会转化为顶礼膜拜的敬意。完美的品质、极致的服务更是酒店尊贵的内涵。最优化的流线设计，可以让贵客享受独一无二的宁静和安全感；高科技智能化在客房服务中的完美融入，给宾客以无所不在的细致关怀。尊贵座驾的贴心配备、套内宽大的观景泳池、一站式商务谈判签约服务……某种意义上昭示着一种梦想的达成和高贵品质生活的实现。在美梦成真的幻觉之中，"财富"对"尊贵"的敬意更加强烈。

Wealth is the foundation and nobleness is the sublimation. Nobleness would not only simply include material satisfaction, but also further cover culture, taste and quality. So the nobleness would be of even more connotation. "Being from wealth to nobleness until honorableness" is the most fundamental conception in our design of Jiafu Peninsula International Hotel. This is also the strongest psychological expectation of China's current wealth class.

The honorable temperament is derived from the combination of both moral and physical practices. The external appearance is not only the resplendence and magnificence, but also further the attractive momentum. The public space of the hotel demonstrates the fantastic luxury with the purest European classical style. But such kind of luxury is not the multifarious stack. The clear and smooth verve commands all the luxury elements to make the whole space be not only of elegant and magnificent temperament but also of authority in the commanding position like the imperial palace. When you stand still there, you could have the respect to worship transformed from the lofty sentiments to spend gold on one throw and the low key to deliberately hidden wealth.

The perfect quality and superlative service are even more the honorable connotation of the hotel. The optimized streamline design could make the distinguished guests enjoy the unique tranquility and sense of security, and the

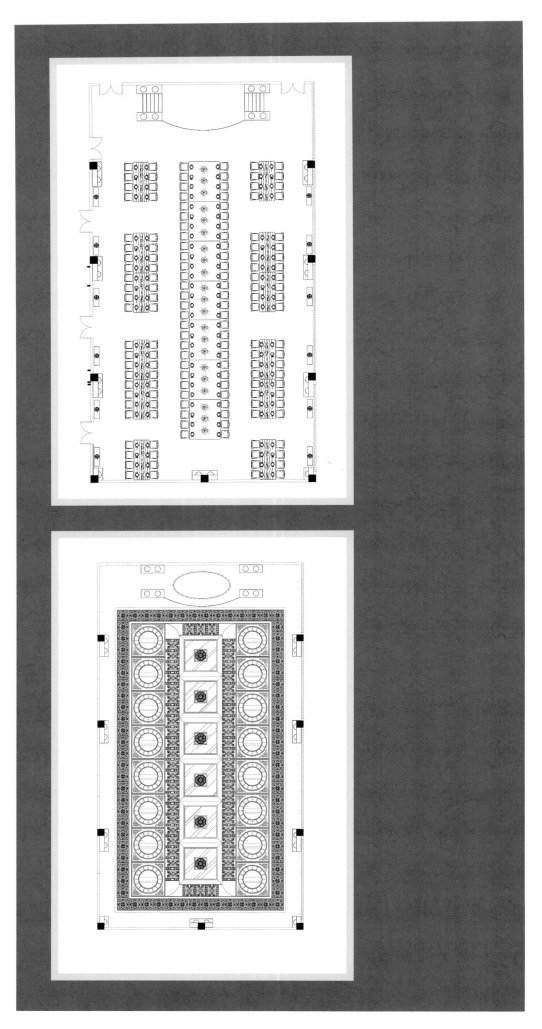

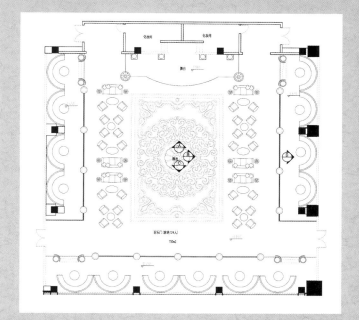

"集人间至味，悟大美之道"是家富半岛国际大酒店打造尊贵的又一利器。气吞山河的美景、中西合璧的美味、才貌具佳的美女，更冲击和浸润着所有宾客的感观和心灵，品位在不经意间悄然升华。

百乐门，曾经演绎着20世纪30年代的上海情调，虽纸醉金迷但不失优雅，矛盾在此以一种暧昧的方式和谐地统一在一起。因而这种情怀，虽历沧桑却仍然撩拨着无数人的心弦。今日的家富国际大酒店以一种更为尊贵的方式演绎着经典的格调。美丽、智慧、独立而充满个性的女人，散发着优雅而深刻的性感和高贵，摄人心魄，给人以无限的想象，却又因难以驾驭而不敢轻易尝试，这种又青睐又畏惧，便是另一种尊贵的修炼。

perfect integration of high-tech intellectualization in the room service to render omnipresent and meticulous care. The intimate equipment of honorable cars, the ample viewing swimming pool inside the room, one-stop service for business negotiations and signing of contracts… All of these declare, in a sense, the effectuation of dreams and the realization of noble-quality life. In the hallucination to realize fond dreams, there would be an even more intense respect of "wealth" to "honorability".

The "collection of the excellent tastes in human world and perception of the doctrine of great beauty" is another sharp weapon for the Jiafu Peninsula International Hotel to build honorableness. The majestic and beautiful scenery, the Chinese and Western combined delicacy and the belles endowed with both beauty and talent would also impact and infiltrate the perception and spirit of all the guests, in which their taste would be quietly sublimated with no notice at all.

The Paramount Hall once deduced the old Shanghai sentiments in 1930s, where although it was a luxury and dissipative life there was still elegance, when the contradictions there were integrated in an ambiguous and harmonious way. Therefore,

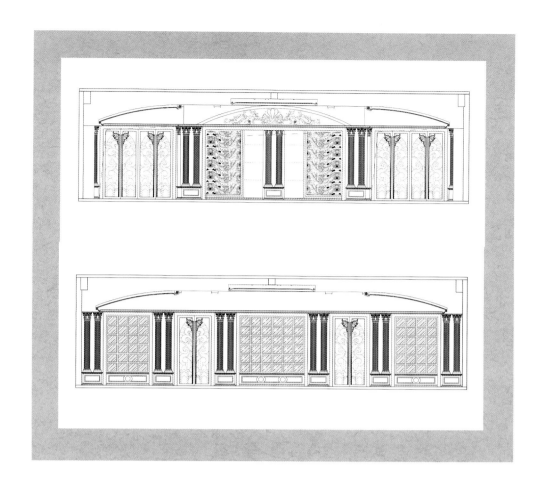

尊贵，可以超越世俗，可以超越浮华，因而可以恒久。
尊贵，是需要修炼的。
在此，财富将在引领下逐步得到精神和品位的升华。
当财富能够泰然处之，便会生出一种皇者的典雅，以及内敛的霸气，这就成就了尊贵。
而家富国际大酒店，以对品质尽善尽美的不懈追求，终将修炼成为中国公认的贵族品牌。她，将是尊贵的代名词。

although such a sentiment has experienced historical storm and stress for quite some years, it is still there to provoke the heartstrings of innumerable people. Today's Jiafu Peninsula International Hotel is deducing the classical style in an even more honorable way. Those women are beautiful, intelligent, independent and full of personality. They are there to send forth elegant and profound sexuality and nobleness to set senses afire. They give you the infinite imagination, but you dare not easily to try it because your difficulty to control them. Such a mentality of both favor and fear is just another cultivation of honorableness.

Since the honorableness could transcend the common customs and the vanity, it would be eternal.

The honorableness needs to be exercised. Here, the wealth would be commanded to make a gradual sublimation in respect of spirit and grade. When you could treat wealth in a calm way, you would produce an imperial elegance and an implicit domination. This makes the honorableness.

With her untiring pursuance for the perfect quality, the Jiafu Peninsula International Hotel would eventually cultivate herself as China's recognized noble brand. She would become the synonym of honorableness.

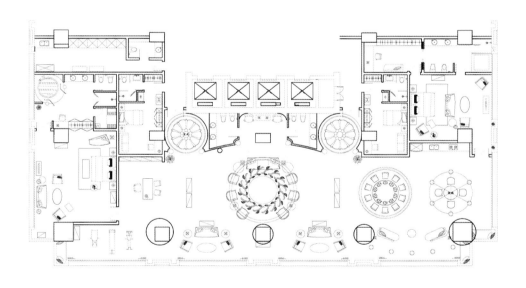

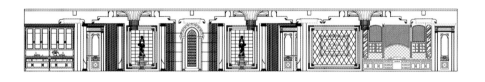

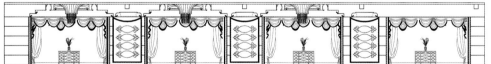

大观园酒店
Grand View
Garden Hotel

项目名称 _ **大观园酒店**	Project Name_**Grand View Garden Hotel**
项目类别 _ **酒店**	Project Category_ **Hotel**
设计面积 _ **30 000 m²**	Design Area_ **30,000 m²**
项目地点 _ **北京**	Project Location_**Beijing**

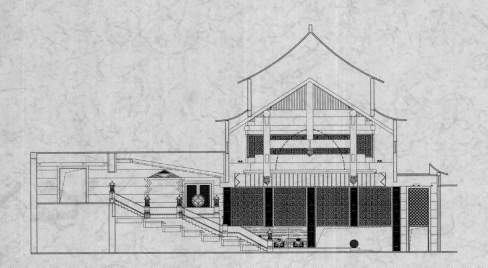
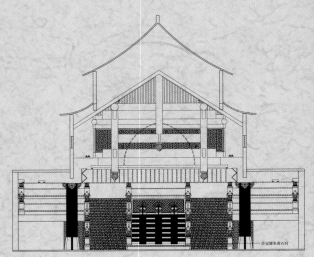

《红楼梦》是中国文化的代表，是民族哲思慧性的档记。大观园更是"合众妙为一妙"的红楼文化的缩影。若说红楼文化元素，则万人有万言，只有"博大精深"一词可论之。红学泰斗周汝昌认为大观园以水为命脉，水是生命之源头，是智慧之底蕴，是美之显相。因而有诗曰："红楼文化水居先，秀色灵情气最鲜；解得鸿濛原蕴水，深情似水水如天。"只有水，具有这等容量，可容得下中华数千年的泱泱文化，可容得下世间的千姿百态，才可容得下红楼大观的真幻与美丑。因而，我们的设计，便循着"水"这条命脉去参悟、去创作、去呈现这红楼一梦的大观之境。

当宾客步入酒店大堂，第一感受便是稳重富贵的太平气象。设计师首先以大章法、大手笔的建筑语言去诠释磅礴的大观之势，"繁华地、富贵乡"正是大观园诞生的背景。

"A Dream of Red Mansions" is the representative of the Chinese culture and the documentary file of the national philosophizing and intelligence. The Grand View Garden is the epitome of the Red Mansion Culture, which is the "combination of multiple fantasies into one". In respect to the elements of the Red Mansion Culture, one million people would have one million explanations, but they could be epitomized as a phrase of "being broad and profound". Mr. Zhou Ruchang, the master of "Redology", holds that water is the lifeblood of the Grand View Garden, since water is the source of life, the deposit of intelligence and the appearance of beauty. Therefore, a poem says, "Water takes precedence in the Red Mansion Culture, and spirit keeps the freshest in the charming beauties and their spiritual passions; if you understand the original deposit of the vast and misty water, you would find passions as deep as water and water like heaven." Nothing but water could have such a capacity to hold the Chinese magnificent culture with a history of thousands of years, to hold the different poses and expressions in the human world, and to hold the reality and fantasy, beauty and ugliness in the grand view of the Red Mansions.

Therefore, our design would follow "water", the lifeblood as mentioned, to perceive, create and present the Grand View of such A Dream of Red Mansions.

When the guests enter into the hotel lobby, their first feeling is the peaceful atmosphere of modesty and fortune. In the first place, the designer uses the architectural language of great technique and master stroke in interpreting the majestic momentum of the Grand View, and the "flourishing land and rich territory" is just the background to create the Grand View Garden.

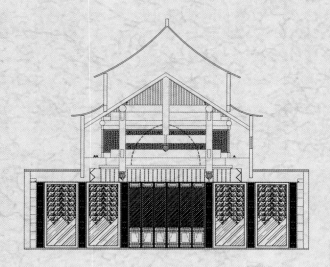
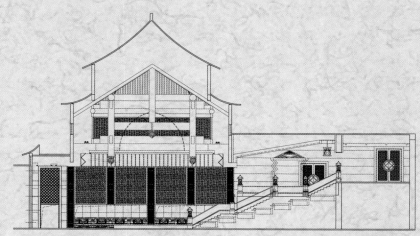

然而,《红楼梦》并非止于漠漠的现实尘境,更呈现出一种清灵的理想幻境,这幻境便是贾宝玉与众女儿的桃花源。设计师洒脱自如地以空间的层次、色调的对比表现出了矛盾交织的大观园。

幻花鼓屏风之后,超然于富丽堂皇之上的,是一脉源自婀娜女儿的清泉净流。女儿是水做的,风流飘逸的黛玉,稳重端雅的宝钗,大气潇洒的湘云……大观园众女儿清洁的品格、芬芳的性灵、超凡脱俗的风貌、丰逸灵动的才思文笔皆如水一般清灵委婉。而女儿们的青春生命、柔情愁绪,更融入了那一脉清澄洁净、含香沁芳的泉流之中。正是因为晶莹如水的女儿,大观园成为荣宁二府中独有的别尘仙境。今日若宾客静立于水边或缘水而上,眼前不禁会上演宝钗扑蝶、湘云醉眠、晴雯撕扇等曼妙佳境。这已不是建筑形式,而是设计师用性灵凝化而成的文化雕塑,弥足珍贵。

However, A Dream of Red Mansions is not confined to the realistic world of a vast expanse, but further assumes the ideal dreamland of pure spirit, and such a dreamland is the Peach Garden of Jia Baoyu and various girls. The designer uses, in a free and easy way, the spatial gradation and tonal contrast in demonstrating the Grand View Garden in interweaved contradictions.

Behind the screen of fantasy flower drum, there is a clear spring of net flow derived from graceful girls, which is transcendental to the magnificence. Girls are made of water. The romantic and graceful Daiyu, the steady and elegant Baochai, the generous and smart Xiangyun… The clear character, the aromatic spirit, the extraordinary appearance, the graceful creativeness and smart writing of all the girls in the Grand View Garden are all clear, smart and euphemistic like water. And the youth and life, gentleness and melancholy of the girls are blended into the flowing fountain that is lucid and clean, fragrant and aromatic. It is just because of the girls who are crystal and translucent like water that the Grand View Garden could become the unique world and fairy land in both Rongguo House and Ningguo House. If the guests today stand at the waterside or stroll along the water, they cannot help but producing such lithe and graceful scenes as Baochai catching butterfly, Xiangyun sleeping drunk and Qingwen tearing fan. This is no longer the architectural form, but a cultural sculpture coagulated of spirit by the designer, which would be very precious.

为了将真幻时空的大开阖展现到极致,设计师更别具匠心地利用最新全息成像技术呈现"黛玉葬花"及"花落水流红"的绝美幻境:泪水淋漓的水芙蓉黛玉楚楚动人摇曳于水中央,幽隐的痴与泪,随飘花落入沁芳的水流,宾客将在梦幻与现实互往中反复品味,前人事,眼前景,心中情,缠绵悲慨,亦真亦幻。

在此有限的视界中,设计师更在人们的心里构建了一个无限的空间,在短暂中获得永恒,这是一种契机。

In order to reveal the great opening and closing of the realistic and fantasy space-time to the utmost, the designer uses, with his own originality, the latest holographic imaging technique in creating the wonderful dreamland of "Daiyu burying flower" and "flowers falling down and water flowing red": The beautiful tear-shedding water lotus Daiyu is swaying in the centre of water, and her shady infatuation and tears fall down into the fragrant water with the floating flowers…The guests would repeatedly savor the taste in the interaction between fantasy and reality. The historical story, the current scene, the emotional heart and the lingering sadness are all either realistic or fantasy.

In the limited field of view, the designer has built an infinite space in human's mind to get a momentary eternity. This is an opportunity.

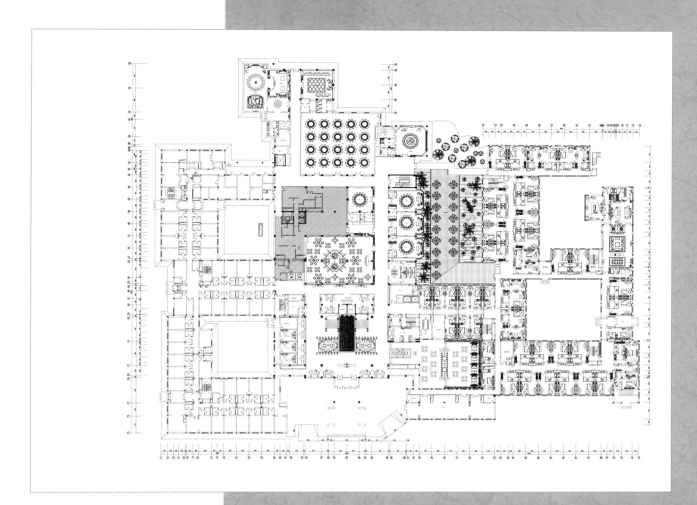
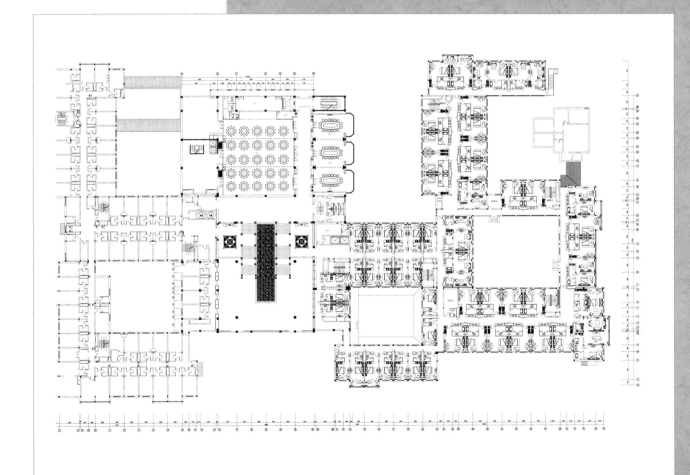

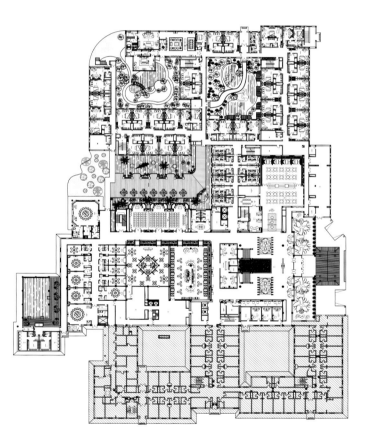

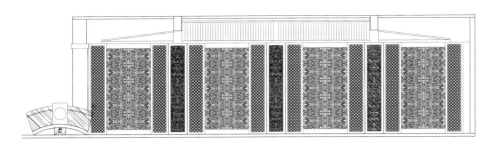

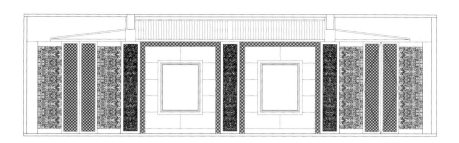

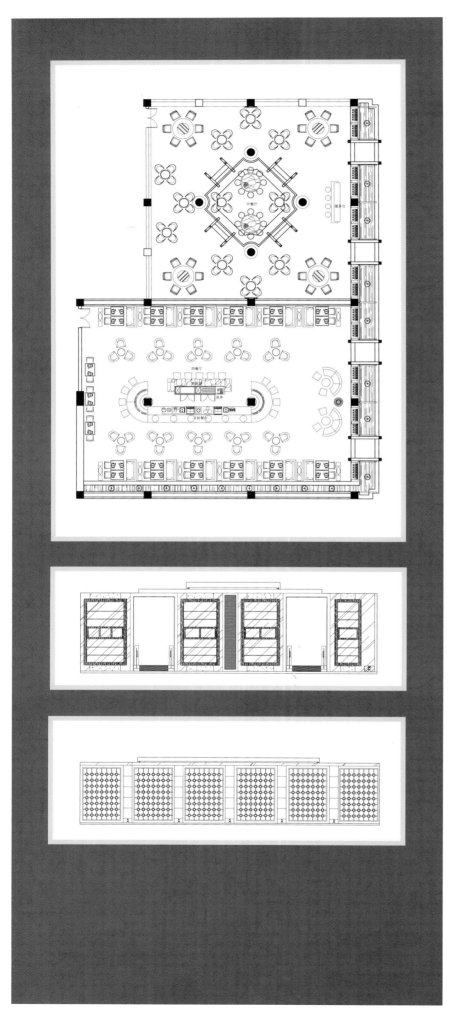

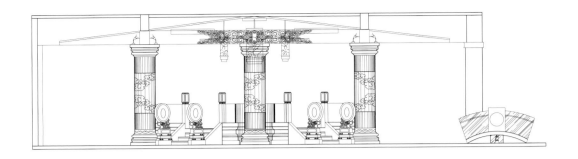

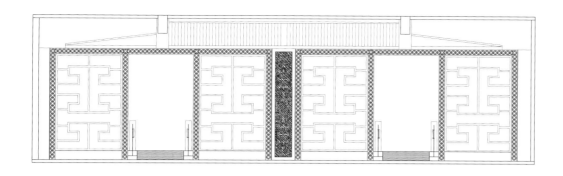

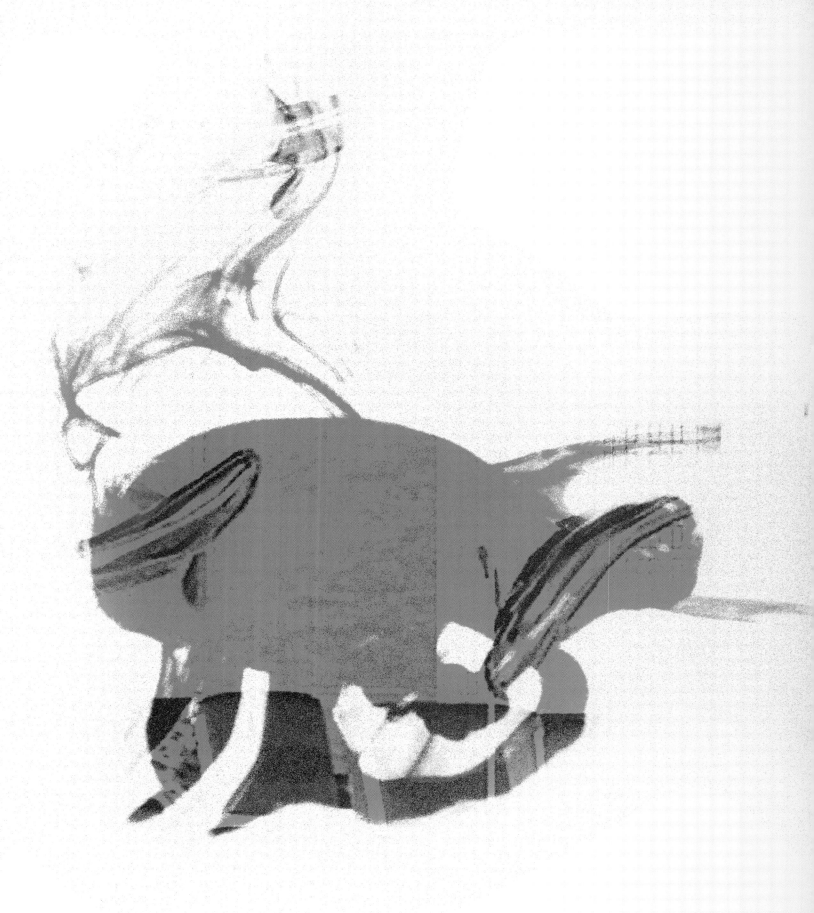

6号院酒店
No. 6 Courtyard Hotel

项目名称_ **6号院酒店**
项目类别_ **酒店**
设计面积_ **10 000 m²**
项目地点_ **北京**

Project Name_ **No. 6 Courtyard Hotel**
Project Category_ **Hotel**
Design Area_ **10,000 m²**
Project Location_ **Beijing**

在北京建一座现代简约风格的酒店并不难,但要在华北平原上构建起"天高任鸟飞,海阔凭鱼跃"的自由空间却并非易事。质朴清新,通透而俊朗,灵动而富于变化,这就是6号院酒店所呈现出的独特气质。

大堂正中环抱的双梯,安宁祥和,以一种欢迎的姿态迎接八方宾客,构建了整个酒店沉稳和谐的氛围。大厅顶部的球形穹顶,映现着深蓝色的夜空。从天窗倾泻而下的水晶灯珠,如甘霖润泽大地,将天与地奇妙地接合,加之中间象征人类精神的主题雕塑,构成了"天、地、人"在物质与精神上融合互生的意境空间。

无论是宴会厅波浪式吊顶,还是餐厅长廊的"小桥流水人家",抑或是大厅的群鱼"竞游",都带来云水的飘逸和流动。人在这样的氛围里,心情也随之荡漾起来,浮动在碧波绿藻之间的是一种自然清灵的情怀。

It is not difficult to build, in Beijing, a hotel of the modern and concise style, but it is really not easy to construct, in North China Plain, a free space that is "as high as the sky for the birds to fly and as broad as the sea for the fish to spring". Naivety and freshness, transparence and smartness, and cleverness and changeability are the unique temperaments for this Hotel to demonstrate.

The dual escalators surrounded in the middle of the lobby are tranquil and peaceful in a posture of welcome to meet all the guests from all directions. This constructs the sedate and harmonious atmosphere of the entire hotel. The spherical vault on the top of the lobby reflects the night sky of dark blue color. The beads of the crystal light pour down from the scuttle, which, like the timely rainfall to moisten the earth, has made a wonderful joint between heaven and earth. With the thematic sculpture as the symbol of the human spirit located in the middle, it has created such a space of realm, in which "Heaven, Earth and Humanity" are blended together for commensalism both physically and mentally.

对于就餐本身，就像观赏一场视觉与味觉的大片，设计师有意在有限的空间中预演一次视觉盛宴。通过对"正方体"这一单体元素空间维度和质感维度的变换与罗列，力求营造一个空灵、纯粹的就餐环境。天花板部分的设计，与普通的餐厅（宴会厅）不同，抛弃了千篇一律的"藻井式"，继续延续"正方体"的构成元素，通过高低错落、此起彼伏的流线造型，巧妙掌控着客人与服务区的空间尺度。"方体"转换成"方形"，天花延续至墙面，浑然天成，演绎着"方"的变奏曲。酒店客房的设计充分体现出人文色彩。以舒适、简洁、通透、典雅的设计理念将狭小压抑的"临时空间"转变为小巧雅致、亲切宜人的"私人住所"，一脉卷轴画卷，瞬间便使心灵完成了从"神色匆忙"向"淡泊宁静"的转变。一种清新隽永的情怀，身住心也驻，这正是现代匆匆旅者所希冀的休憩环境。

融天、地、人于一体，这就是整个6号院酒店的胸怀和品性。而细节的处理，精致细腻却又干净简练，自然气息无处不在。这是人与自然、天与地、水和木的和谐共处，创设出的奇妙的混搭情境，让每一个接近它的人都流连忘返、心意相属。

No matter it is the wave-mode suspended ceilings of the banquet hall, or the "tiny bridge, flowing brook and hamlet homes" in the restaurant corridor, or the "competing swimming" of a group of fish in the lobby, all of them have brought about the elegance and motion of clouds and water. In such an atmosphere, the human feelings are rippling therewith, and what floats in between the green waves and green algae is the natural, pure and smart feeling.

As for the dining itself, it should be something like watching a blockbuster of visual and gustatory senses. It is the designer's intention to preview a visual feast in a limited space. Through the transformation and enumeration of such a monomer of "cube" in its spatial dimension and textural dimension, he would seek to create an intangible and pure dining environment. The design of the ceilings is, in fact, different from that of an ordinary restaurant (banquet hall), in which he abandoned the monotonous "shaft sinking", but continued to use the structuring element of the "cube". In doing so, he made a clever control of the spatial dimension of both the guests and the service zone through the streamline modeling of scattering at random discretion and ranging up and down. As the "cube" is converted into "square", the ceilings are extended to the wall space with the highest quality to deduce the "variations of cube".

The design of the hotel guest rooms fully reflects the humanistic color. With the design concept of comfort,

conciseness, transparence and elegance, the narrow and depressive "temporary space" is converted into a small, refined, cordial and pleasant "private residence". A scroll of painting could fulfill, in a moment, the spiritual transition from the "hasty look" to "simple tranquility". The clear, fresh and pleasant feelings exist not only in the body but also in the heart. This is just the environment of recreation for the modern hasty travelers to expect.

It is the bosom breast and moral character of the entire No. 6 Courtyard Hotel to melt Heaven, Earth and Humanity into one melting furnace. However, in the processing of details, we could see the delicacy and exquisiteness but cleanness and conciseness as well as the natural breath everywhere. This is the harmonious coexistence of humanity and nature, heaven and earth, and water and wood to create a wonderful and mixed situation, in which each human being close to it would enjoy himself so much as to forget to leave with his emotional involvement.

德州国宾馆

Dezhou State Guesthouse

项目名称 _ **德州国宾馆**
项目类别 _ **酒店**
设计面积 _ **18 000 m²**
项目地点 _ **山东德州**

Project Name_ **Dezhou State Guesthouse**
Project Category_ **Hotel**
Design Area_ **18,000 m²**
Project Location_ **Dezhou, Shandong Province**

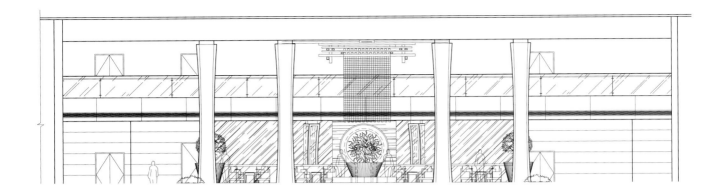

德州国宾馆，是坐落于"董子读书台"旁的文化主题酒店。文化的主题确定为鸿儒董仲舒的思想之火，以及汉代的文化艺术之光。

《天人三策》是董仲舒开启中华两千年儒学治世历史的旷世神篇。以古篆镌刻于木简之上的《天人三策》铺满酒店大厅的天花，一脉而下延至总服务台，使得每一个南来北往的客人都会在第一时间注意到它。虽然其文字和涵义鲜有人知，但在驻足凝望之际，每一个人都不禁会发怀古之幽思，徜徉于中国历史文化的长河。

大厅正中赫然悬挂着一个巨大的活字印刷机造型，它昭示着中华文化千年不断的传播与传承。它虽静默于此，但伫立其下的每一个华夏子孙都可以听到，它在以洪钟之音诵读着两千年前的圣贤之言，永不停息。

大厅中水池清灵泉涌不绝，有源之水生生不息，象征着董子的思想源远流长。水是中国哲思的源泉，无论是道家的"上善若水"，还是儒家的"水有五德"，水都是智慧的象征，因而君子遇水必观。立于水边，水天映人，天人合一。

大厅中的浮雕造形柱，创意大胆，极富内涵。书的造型令通天彻地的柱子顿生灵慧，再配以气韵奔放的云纹，既严谨古朴，又飘逸浪漫，动静相宜，充分体现了儒家和谐一体的大一统思想。

汉代艺术装饰风格，可以用"质"、"动"、"紧"、"味"四个字来概括，即古拙质朴与浪漫的气质，流动不绝的节奏与韵律，形式的饱满与均衡，整体格局的对比和统一。在每个空间和细节的设计中，我们充分把握这些特质，巧妙地应用灯具、家具、艺术陈设等元素

Dezhou State Guesthouse is a hotel of cultural theme, located near the "Dongzi Reading Table". The cultural theme is determined as the ideological spark of the great scholar Dong Zhongshu and the cultural and art light of Han Dynasty.

The "Tian Ren San Ce (Three Policies on Heaven and Humanity)" was the outstanding fairy articles for Dong Zhongshu to open the Chinese history of governing the country by Confucianism for two thousands of years. The "Tian Ren San Ce (Three Policies on Heaven and Humanity)" engraved in seal characters on wooden slips is decorated on the ceilings of the hotel lobby, with its range extended to the front desk, so as to make all the guests from all over the country could notice it at the first time. Although few people know the characters and meanings thereof, yet on the occasion of their standing there to read it, each of the guests cannot help producing the meditation on the past and strolling in the long river of the Chinese historical culture.

Impressively hung in the middle of the lobby is the huge model of a type printing press, which shows the constant dissemination and inheritance of Chinese culture for thousands of years. Although it is silent there, each of the descendants of the Chinese nation could hear it reading, in a voice like a large bell, the words of sages of two thousand of years ago with no interruption.

The water pool in the lobby is clear and smart, with unceasing gush of springs. The water with good fountainhead flows in an endless succession, which symbolizes the long standing of Dongzi ideology. Water is the source of the Chinese philosophizing. No matter it is the Taoist "As Noble as Water" or the Confucian "Five Virtues in Water", water is the symbol of wisdom. Therefore, whenever water is in sight, a man of noble character would be there to observe it. When you stand at the waterside, you are being reflected by both water and heaven, when you could feel the unity of heaven and humanity.

In the lobby, the anaglyph-shaped column is of bold creation and rich connotation. The book sculpture makes the extremely high column instantly producing wisdom. As it is further matched with the bold artistic conception clouding, it looks not only rigorous and quaint, but also elegant and romantic. The suitability of motion and stillness fully embodies the Confucian ideology of grand unification of harmonious integration.

The artistic adornment style in Han Dynasty could be summarized as four words of "quality", "motion", "tightness" and "taste", which are, in details, the classic, simple and romantic quality and temperament, the unceasing flow and motion of rhythm and cadence, the full and balanced form and the contrast and unification of the entire structure. In the design of each of the spaces and details, we make a full control of these peculiarities, and cleverly use such elements as lamp, furniture and art display to enhance atmosphere, so as to create a modest but colorful lobby, elegant and honorable restaurant, and warm and comfortable guest rooms to reflect the artistic charm and noble momentum of Han Dynasty at the utmost.

烘托氛围，从而营造出稳重而华美的大厅，高雅而尊贵的餐厅，温馨而舒适的客房，最大限度地彰显出汉代艺术的韵味和尊贵之气。

汉代是我国历史上一个大融合的时代。汉代艺术是建立在对各种艺术的包容、吸收和融合基础之上的，因而具有博大与辉煌的特质。传承这种特质，我们在设计时不拘于任何形式，不守旧，着意于创新，将文化、品位、高雅、庄重、华贵融为一体，既渲染东方文明又不失西式现代风格，在设计上达成了东西方、传统和现代的和谐与统一。

In the Chinese history, Han Dynasty was an era of grand integration. The art of Han Dynasty was built on the basis of inclusion, absorption and integration of various arts, and it was, therefore, of broad and brilliant peculiarity. In our inheriting these characteristics, we do not confine ourselves to any form whatsoever in the design. And we are not conservative, but we attempt to make innovations to integrate culture, taste, elegance, solemnity and luxury into one entirety, in which we have not only rendered the Oriental civilization but also adopted the Western modern style, so as to realize the harmony and unification of the Oriental and Western styles, and the traditional and modern conceptions in our design.

天和国际大酒店
Tianhe International Grand Hotel

项目名称 _ **天和国际大酒店**
项目类别 _ **酒店**
设计面积 _ **22 000 m²**
项目地点 _ **重庆**

Project Name_ **Tianhe International Grand Hotel**
Project Category_ **Hotel**
Design Area_ **22,000 m²**
Project Location_ **Chongqing**

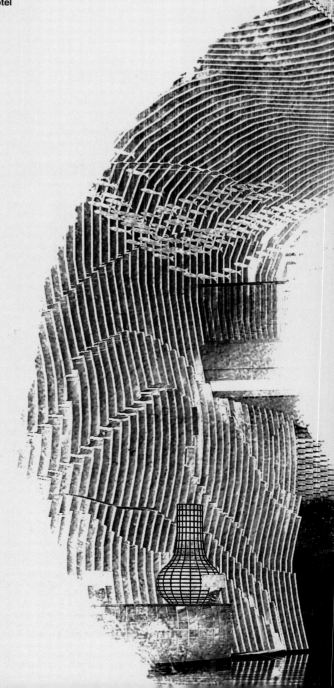

天和国际大酒店坐落于重庆鱼洞江畔，呈长江之灵气，秉云篆山之柔美，引水中之鱼的欢畅带着"女娲补天，瀑挂龙鳞"的美丽神话走向我们。设计师以破传统、超现代的理念，通过搭建视觉与理性的冲突，成就天和的聚焦。

铸就一次个性舒适、绚烂浪漫的独特体验是本案设计师最希望为客人打造的。透过纯粹的装饰、纯粹的矛盾、纯粹的活力，设计师向我们展现了一幅炫尚温馨的酒店图景。极致的追求，灵动的空间，致使酒店已不仅是住所，更是放空大脑、静享沉思、汲取灵感的绝佳

Tianhe International Grand Hotel is located on the Yudong riverside, which shows the spirit of Yangtze River and holds the tenderness of the Yunzhuanshan Mountain. It comes to us with the beautiful myths of "Goddess Nuwa Repairing the Sky" and "Waterfalls Flying on Dragon Scales". With his tradition-breaking and ultramodern concept, the designer has made a success in focusing the Tianhe Hotel through constructing the conflict between visual sense and rationality.

It is the greatest hope for the designer of this case to build, for the guests, a unique experience of personal comfort and splendid romance. Through the pure decoration, pure contradiction and pure vigor, the designer shows us a colorful, fashionable and warm picture of the hotel. The superlative pursuit and smart space have made the hotel not only a residence but more also the superexcellent place to empty the minds, enjoy meditation and get inspiration.

场所。

棕褐色的地板、碧蓝色的天顶，酒店大堂于稳重中透着灵气。厅堂中央的那一潭流水在四棵棕榈树的陪衬下，使原本空旷有余的共享空间顿时热闹了起来。一动一静、一水一树的搭配，既成室外一景，又与室内空间浑然天成。

雾都重庆，向来是闲适与享受的合集。天和酒店虽为商务酒店，但经设计师的巧手雕琢，不仅毫无刻板单调之感，反具浪漫时尚的绝佳住所，给予住客充分完全的放松与舒适。忙碌之余，于此厅此室中，泡上一杯咖啡，享受一次沐浴，静观一次夜景，看一江春水向东流，该是怎样一种闲适！

With dark brown floor and dark blue ceiling, the hotel lobby looks smart but modest. The pond of flowing water in the middle of the lobby is accompanied by four palm trees, which would immediately make the originally empty and superabundant shared space filled with a lively atmosphere. The collocation of motion and stillness and that of water and trees have made it not only become an outdoor scene, but also integrate with the interior space so naturally.

Foggy Chongqing is always the integration of leisure and enjoyment. Although Tianhe Hotel is a business hotel, with skillful carving of the designer, it has been made such a superexcellent residence as being not only without inflexible monotone but of romantic fashion to render a full and complete relaxation and comfort to the guests. When you come to the Hotel after your busy work, to drink a cup of coffee, to enjoy a bath, to contemplate the night view or to watch a river of spring water rolling towards the east, what a leisure and comfort it would be!

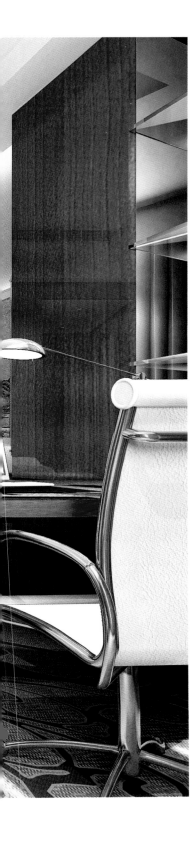

首钢酒店
Shougang Hotel

项目名称 _ **首钢酒店**	Project Name_**Shougang Hotel**
项目类别 _ **酒店**	Project Category_ **Hotel**
设计面积 _ **50 000 m²**	Design Area_ **50,000 m²**
项目地点 _ **北京**	Project Location_**Beijing**

首钢酒店意在营造一个高耸阔大的垂直空间，展现首钢人气魄冲天的凌云壮志，这是对首钢文化的绝妙诠释。

酒店大堂极尽高耸之所能，四周墙壁宛如铜墙铁壁，升腾起无可匹敌的力量。然而屋顶却变为透明，与天空相接，直冲九霄的是首钢人勃发的激情和干劲。大堂正中，三座白色圣火台柱燃着橙红的"火焰"，它是首钢的生命之火，是首钢人自强不息的信念。如此这般的力量感已给人以"刚硬"的震撼，然而大堂中央葱葱郁郁的绿意，将自然生态的气息引入，化刚毅为柔韧。虽不胜绚烂，却已燃起清新的惬意。人于此则亦可刚亦可柔，被宏大感染，也为温婉浸润。设计师在这里，传达出的自然与人文的和谐统一是建筑的最大亮点。

Shougang Hotel is intended to create a towering and broad vertical space to demonstrate the lofty aspirations of Shougang people with towering boldness to reach the clouds. This is the wonderful interpretation of Shougang's culture.

The Hotel lobby is made as towering as possible with the all-side high walls built as impregnable fortress, so as to create an unapproachable strength. However, the roof is made transparent to kiss the sky to show that it is the vigorous passion and drive of Shougang people to reach as high as into the space. In the middle of the lobby, there are three sets of white Holy Flame stands. The burning flame of orange red color is the fire of Shougang's life and the self-improvement conviction of Shougang people.

Such kind of sense of strength is of great shock with "rigidity", but the exuberant greenness in the middle of the lobby has introduced the ecological breath of nature to turn the fortitude into flexibility. Although it is not splendid enough, yet it has already produced the fresh satisfaction. The human beings here could be either rigid or flexible, since they have been either impacted by the magnificence or infiltrated by the gentleness.

It should be the greatest light spot of the architecture for the designer here to convey the harmony and communion between nature and humanity.

燕都酒店

Yandu Hotel

项目名称 _ **燕都酒店**	Project Name_**Yandu Hotel**
项目类别 _ **酒店**	Project Category_ **Hotel**
设计面积 _**50 000 m²**	Design Area_ **50,000 m²**
项目地点 _ **河北石家庄**	Project Location_**Shijiazhuang, Hebei Province**

伏羲亘古的传说、正定古城的灿烂、秦皇古道的苍凉、赵州城头变幻的风霜……古城石家庄，今已蜕变重生。富丽奢华或许不属于她，然而她却独拥"燕都"的高贵和格调。燕都酒店高贵而不招摇，独具格调却又亲近可人。大堂的八棵椰树将时空成功转移，恍若置身海南的热带风光，可以欢畅奔跑，可以坐卧嬉游。椰树高耸，与"银河式"屋顶相望相依，拉近了上下垂直距离，刨除了疏离，制造出紧凑的空间。在这"银河"的夜空下，仿佛椰树也息了热辣，转而娴静深沉，成为灿烂星辉下最唯美的风景。然而在这椰林中却又巧置各色沙发，令人在内陆也能足享热带休憩之愉。

"身未动而心已远"。心亦如带着翅膀的天使，追逐理想国而去，双手也因此可触及银河。自然与人文、身体与心灵的统一，瞬间在设计师的匠心独运中得以实现。

The ancient legend on Fu Hsi, the brilliance of the ancient city of Zhengding, the desolateness of Qin Huang Ancient Road, frost changes on top of the city wall of Zhaozhou...The ancient city of Shijiazhuang has changed quite a lot since her revival. Brilliance and luxury might not belong to her, but she alone could enjoy the nobleness and life style of "Yandu (Capital of the ancient Yan State)".

Yandu Hotel is noble but modest, and unique in style but close and friendly. The eight coconut trees in the lobby have made a successful transition of the space-time as if it were in the tropical scene in Hainan, where you could run delightedly and enjoy yourself in various activities. The towering coconut trees and the "galaxy-style" roof watch and depend on each other, so as to shorten the vertical distance, to reduce the alienation and to create a compact space. In the night sky of "galaxy", the coconut trees seem to remove their hot flavor and turn demure and deep, so as to become the most esthetic scenery under the brilliant brightness of stars. However, there are sofas of various kinds artfully placed in such a grove of coconut trees, which makes people can fully enjoy the tropical recreation even in the inland.

"Bodies keep close, but souls go apart". The hearts of human beings are also like the angels with wings that would pursue the "Utopia" and thus touch the "galaxy" with their own hands. In the designer's inventive mind, the communion of nature and humanity, and body and spirit could be realized in an instant.

某会所型酒店
A Club-type Hotel

项目名称 _ **某会所型酒店**	Project Name_**A Club-type Hotel**
项目类别 _ **酒店**	Project Category_ **Hotel**
设计面积 _ **19 000 m²**	Design Area_ **19,000 m²**
项目地点 _ **北京**	Project Location_**Beijing**

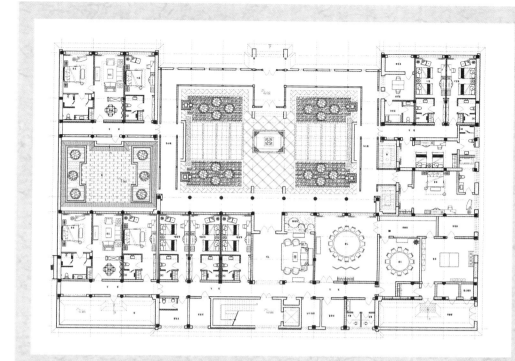

过于现代的设计其实是很容易过时的,这在各个设计领域里都成为不争的事实。所以我们在设计此会所的时候在也风格上作出了不一样的定位。在公共空间以及重要场所的设计都沿用了抽象化的中国古典建筑符号,包括部分空间内陈设的工艺品,体现了古典文明与文化在现代空间中的传承、民族文脉的延续。

为表现出室内装饰的气度与本项目建筑物的气势,在室内装饰特点的体现上,设计中采用大尺度、大体量的形态来构成空间中的主要形态成分,表现出庄重、典雅、大气磅礴的效果,舍弃细小的装饰手段,以体现技术含量较高的箔、烫花艺术玻璃、石材来表达空间的品质和精神。

If a design is too modern, it is in fact quite easy to be outdated. This is an undisputable fact in various design fields. Therefore, when we did the design of this club, we made different style orientations. For the public spaces and important places, we continued to use the abstracted symbols of the Chinese classical architecture in their designs, including the furnishings and craftworks in some of the spaces, to reflect the inheritance of the classic civilization and culture in the modern space and the continuation of the national tradition of culture.

In order to show the tolerance of the interior decoration and the magnificence of the building of this Project, in the reflection of the characteristics of the interior decoration, the design adopted a large dimension and large scale of forms to construct the main forms and compositions in the space to demonstrate the result of solemn, elegance and tremendousness but abandon the slight decorative means, so as to embody the quality and spirit of the space by use of the foil, pyrographic art glass and stone of high technical content.

目前有股风潮,即西方现代建筑思想指引中国传统手法去表现的趋势日渐明显,使人感受到一个最终的共同探究目标:"什么才是中国的现代性?"对于诸多疑问,在该会所的设计成果里找到些许的答案。大面的实木花格、质地光润的米黄石材、配以舒适的古典欧式家具,无处不体现对中国博大精深的传统文化精神进行的一种提炼和延续,融汇更多中国传统国学的精髓,两者结合营造出低调奢华的氛围,使其迸发出奢华的宁静气韵,蔓延到整个空间。

Currently, there is such a trend, in which it is more and more obvious to use the Western modern architectural ideology to guide the Chinese traditional approaches of expression. It makes people feel the necessity to find the eventual goal of common exploration: "What is on earth the Chinese modernity?" As for many questions, you may find some answers from the design results of this club. The solid wood lattice and smooth cream-colored stone used on the bedding face that is matched with comfortable and classical European furniture could fully reflect the refinery and continuation of China's broad and profound traditional spirit of culture, which has blended with even more essence of the Chinese traditional sinology. The combination of both has created a low-key and luxurious atmosphere to ignite an artistic conception of luxury and tranquility that spreads to the whole space.

体现西方审美情趣的元素,将中式的图案加以提炼。该康体项目是度假、休闲型场所,以轻松、典雅来定义主调,周围空间围绕其伸展开来。整体风格延续了建筑本身的中式风格,其中穿插点缀一些体现西方审美情趣的元素。并将中式的图案加以提炼、简化,形成连续性纹样装饰立面,配以挑高的空间尺度,将空间关系衬托得层次格外分明,尽显低调而尊贵的优雅气质。

We have attached great importance to the reflection of the Western aesthetic sentiments and elements and the refinery of the Chinese-type designs. Since this health project is a place for holidays and leisure, the relaxation and elegance should be defined as the main theme, around which the surrounding space would then be extended. The integral style keeps the same as the Chinese style of the main building, but alternated, as adornment, some elements to reflect the Western aesthetic taste. Furthermore, the Chinese-type devices are refined and simplified to form an adornment façade of continuous grain appearance that is matched with the overhanging spatial dimension to make an extremely distinct contrast of the spatial relations to demonstrate the low-key and honorable temperament of elegance.

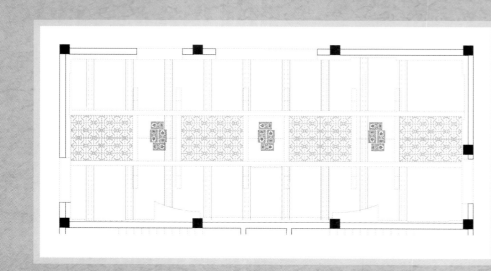

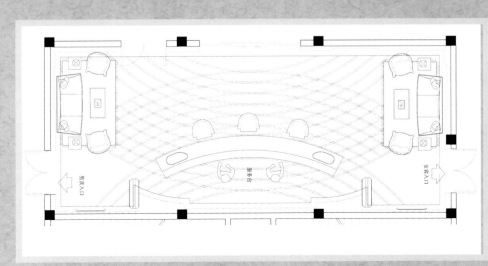

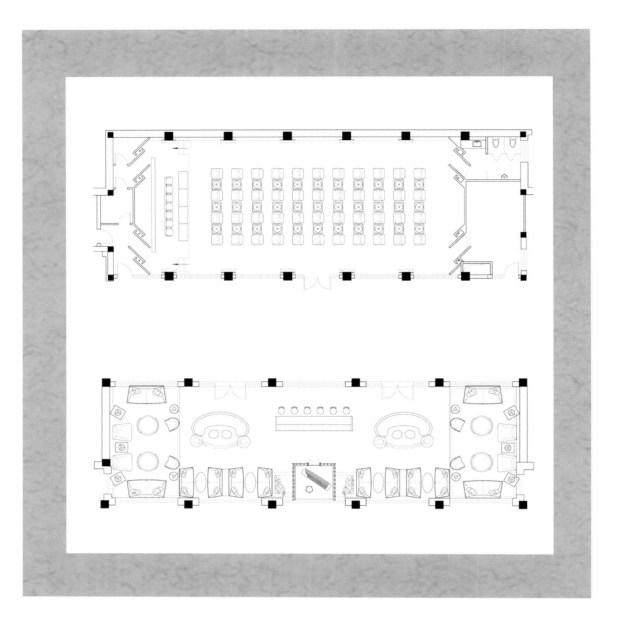

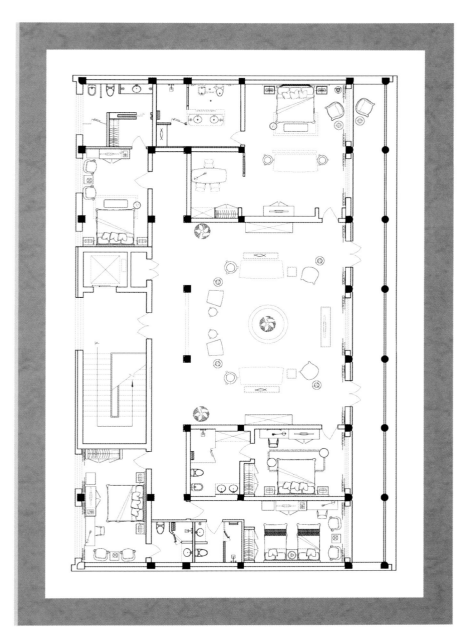

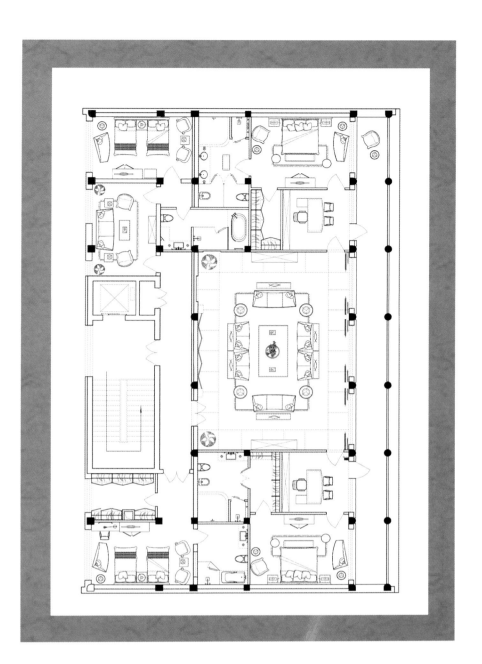

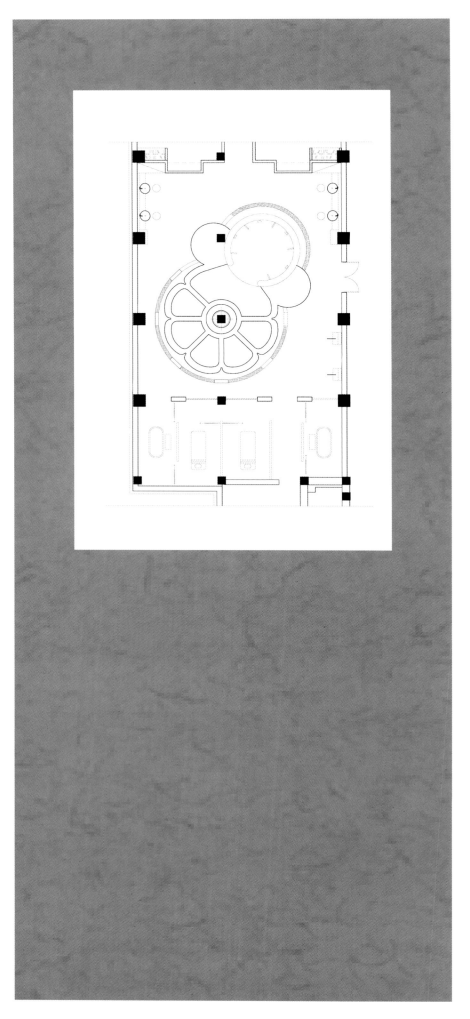

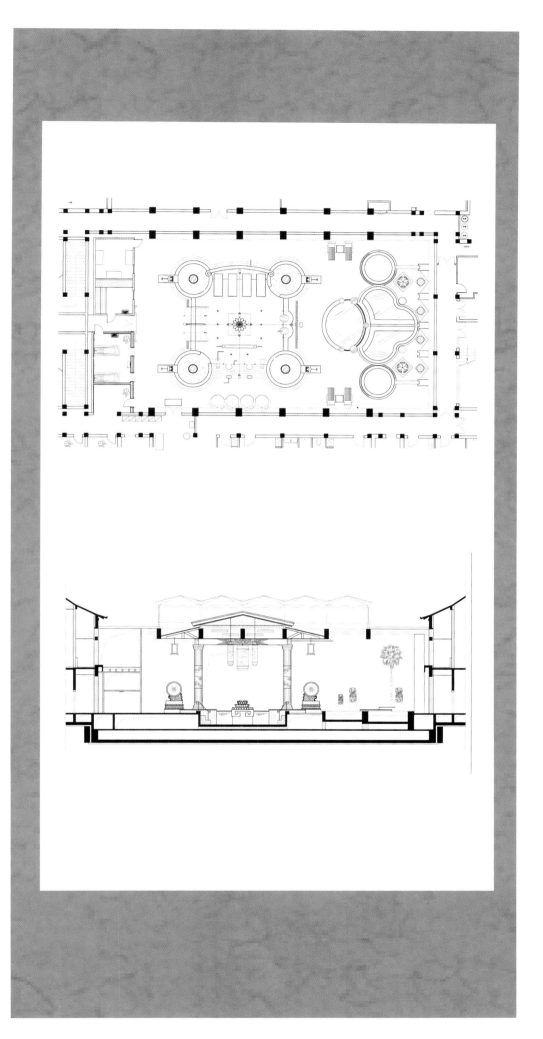

懿赐会所
Yici Club

项目名称 _ 懿赐会所	Project Name_ **Yici Club**
项目类别 _ 会所	Project Category_ **Club**
设计面积 _ **2 000 m²**	Design Area_ **2,000 m²**
项目地点 _ 山西太原	Project Location_ **Taiyuan, Shanxi Province**

这是一位内敛而质朴的晋商委托的项目,他希望构造一个会所空间,既可涤心修炼,又可邀朋引伴,但一定要有文化品位而不落奢华俗套。有文化内涵的会所,在这座以灯红酒绿著称的城市还确实难寻,这种卓尔不群激发了我们极大的创作激情。

深邃富丽的大院,应是山西最有代表性的文化。"由庄外遥望,十数里外犹见,百尺矗立,崔嵬奇伟,足镇山河。"背井离乡苦心经营,一朝业成衣锦还乡。大院是无数走西口的汉子最荣耀的梦,也是无数孤守的女人最坚强的堡垒。大院体现了晋商诚信进取的精神文化,也见证了他们艰苦拼搏的历程,展现了他们纵横商海的辉煌,同时也吟唱着晋商衰败的历史悲歌。

This is a project entrusted by a Shanxi merchant who is introvert and plain. He hopes to construct a club space where the people could not only cleanse their spirits and make their spiritual practice but also invite their friends and companions for entertainment, which would surely be of cultural grade but depart from luxury and convention. In such a city that is famous for the scene of debauchery, it is really very hard to find a club of cultural connotation. Such uniqueness has aroused our great creative enthusiasm.
The deep and splendid courtyard should be the most representative culture in Shanxi. "To look far away outside the village, it could be seen even in a distance of a dozen of kilometers, which stands as tall, elegant, great and peculiar as to suppress the mountains and rivers." In the past, after long time elaboration far away from home, they would return home in glory after they had won great success. The courtyard was the most honorable dream of countless fellow men who left for the west cols and the strongest fortress of myriad solitary women. At

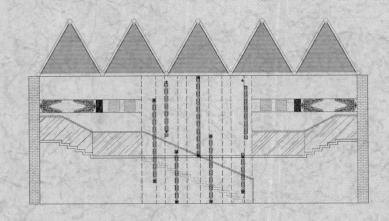

在高大的室内空间构建一个大院式的高墙深宅，设计师在向文化致敬的同时，也为会所主人营造一个修炼心灵的场所。进入会所，迎面便是一组圆型水晶光纤，如明月悬空，映入黑金沙大理石围合的一潭静水，清明透彻。月下之水，刚毅中怀阴柔；水中之月，辉煌中隐悲凄。这大院便映在这水月之中，令人心净如水，沉思如月。大隐隐于市，在纷繁浮躁的世俗之中，却能大智若愚、淡然处之，如莲之与泥共存而不染。如何不能"以史为鉴,再铸辉煌"？在心灵不断的升华中，新一代晋商的凌云之志，便逐渐地蓄积于此。

present, the courtyard reflects the spiritual culture of Shanxi merchants' sincerity and go-aheadism, witnesses their then course of hard struggle and demonstrates the magnificence of their manly toil in the commercial sea, and chants the historical tragedy of the ancient Shanxi merchants in decline.

To construct a courtyard with high walls and deep curtilage in the lofty interior space, while the designer pays his respects to the culture, he would create a place for the club owner to make spiritual practice. When you enter into the club house, you could see a group of circular type crystal optical fiber. It is just like the bright moon hanging in the air to map into the pond of still water surrounded with Black Galaxy marble, which is clear, bright and thorough. The water under the moon is feminine in manliness. The moon on the water conceals the mournfulness in the brilliance. The courtyard is mirrored in the midst of water and moon, which makes the human mind as clear as water and meditation as deep as moon. Great genius often lies concealed, and, in the numerous blundering common customs, he could treat with indifference as if still water runs deep and act as a lotus flower that keeps pollution free even though it co-

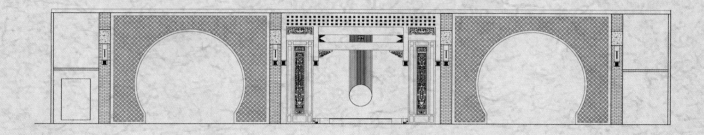

在会所的细节设计中,中华文化"儒、释、道"的精髓以最自然的状态和谐地蕴涵在建筑中。矛盾的统一、情与理的统一、动与静的统一以及人与自然的统一,是和谐人类文明和人性深度的最根本体现。传统与时尚的对接,东方与西方的和鸣,最终融在了中华哲学的大美当中。

水无形而有万形,水无物而能容万物。

建筑构造一种氛围,氛围又影响着人的气息,这就是设计师的责任。

exists with silt. Why can't we "take history as a mirror and recast the brilliance"? In the unceasing spiritual sublimation, the lofty aspirations of the new generation of Shanxi merchants would be gradually accumulated for such a reason.

In the design of the club details, the Chinese culture's essence of "Confucianism, Buddhism and Taoism" is contained in harmony in the building in the most natural state. The unity of contradictions, the unity of feeling and principle, the unity of movement and quietude as well as the unity of human and nature is the most fundamental embodiment to harmonize human civilization and humanistic depth. The docking of tradition with fashion and the acoustic resonance of the Eastern with the Western have been finally blended into the great beauty of the Chinese philosophy.

Water is intangible but could appear in multiple forms, and water is devoid of anything but could contain all things on earth.

Architecture creates a kind of atmosphere, and the atmosphere would influence the human breath. This is just the responsibility of a designer.

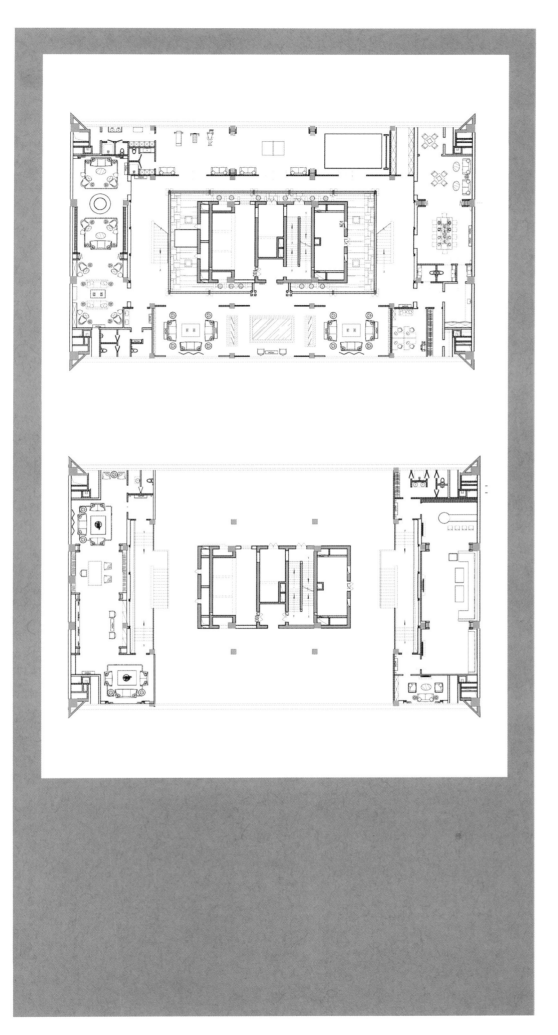

9 号会见厅
No. 9 Reception Hall

项目名称 _ **9号会见厅**	Project Name_ **No. 9 Reception Hall**
项目类别 _ **会所**	Project Category_ **Club**
设计面积 _ **480 m²**	Design Area_ **480 m²**
项目地点 _ **北京**	Project Location_ **Beijing**

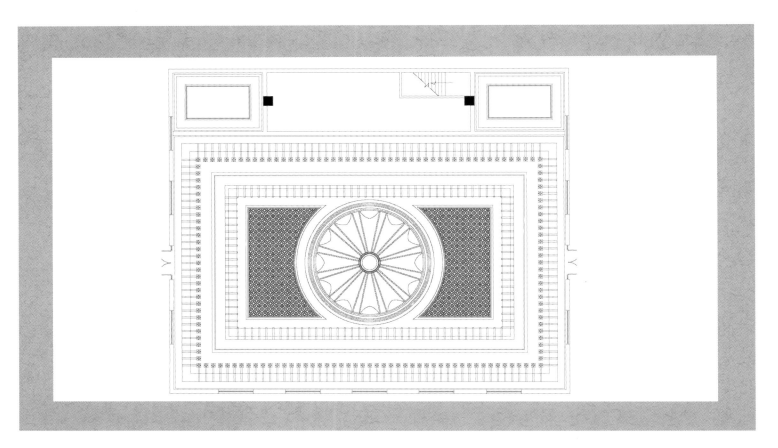

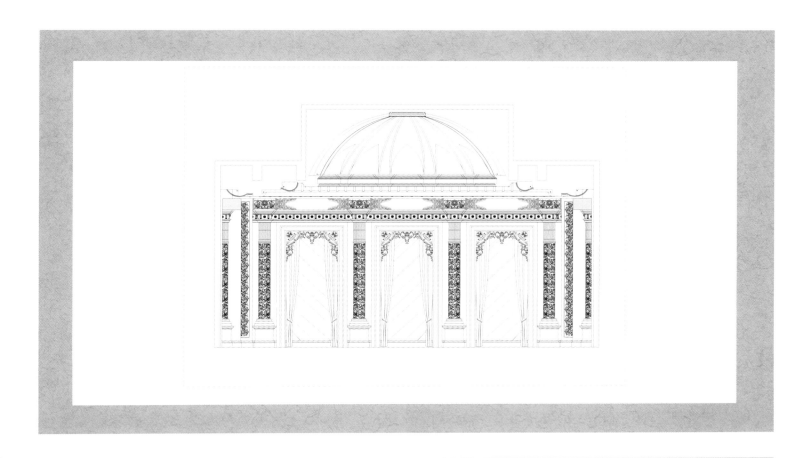

把一个纯粹的空间,构造成气度非凡、格局雄浑、底蕴深厚、意向高远的大和之境实属不易。这需要空间设计师具有同样的品格、胸怀和风范。

在天圆地方的大格局中,一幅气势恢宏的江山图确定了整个空间博大的容量。积淀了数千年的东方文化以坦荡开放的胸怀,将古今中外的万千气象一一融入。在不变中寻求变化,气势连贯,气韵贯通,在雄浑壮阔中实现圆融统一。

致广大而尽精微,于大写意中的细腻、工整、严谨,更有别样的典雅意境。

百川交汇铸就万世伟业,这是东方智慧与精神的完美诠释。

It is really not easy to transform a pure space into a location of grand communion that is of impressive appearance, forceful framework, profound deposit and far-reaching intention. This needs the space designer to have the same character, bosom breast and demeanor.

In the great framework of round sky and square ground, a magnificent drawing of landscape has determined the broad capacity in the entire space. The Oriental culture accumulated for thousands of years has incorporated, with a broad and open mind, all the things of multifarious kinds at all times and in all over the world. It is to seek the transformation from invariability with the consistent momentum and thorough artistic conception, so as to realize the harmony and communion in the forceful magnificence.

It is of another artistic conception of elegance to do the exquisiteness for the broadness and to seek the fineness, carefulness and strictness in the freehand technique.

The intersection of all rivers could create the great cause for ages. This is the perfect interpretation of the Oriental wisdom and spirit.

199 天宫会所
No. 199
Heaven Palace Club

项目名称 _ 199 天宫会所
项目类别 _ 会所
设计面积 _ 3 800 m²
项目地点 _ 江苏徐州

Project Name_ **No. 199 Heaven Palace Club**
Project Category_ **Club**
Design Area_ **3,800 m²**
Project Location_ **Xuzhou, Jiangsu Province**

天宫大厅以深邃的黑为底色，让人们自然而然踏实地步入解脱灵魂的黑夜。对比鲜明的色彩、形状各异的图案、闪烁跃动的光影，随着音乐在和谐律动，以天外玄妙的节奏挑拨着人的神经，抚摸着心底的悸动，随之涌动的脉搏，期盼冲过理性与激情的闸门，心已经飞向一个无边无垠的空间。

The bottom color of the lobby of the Heaven Palace is made as deep black, which makes people step naturally and steadfastly into the dark night to liberate their souls. The contrasting colors, the patterns with different shapes and the jerking and twinkling shadows are moving in harmony with the rhythm of music to provoke the human nerves with the abstruse rhythm from Far Heaven and touch the heart throbbing and the surging pulse, when they look forward to rushing across the gate of rationality and passion, and their hearts have already flown to a space of infinity.

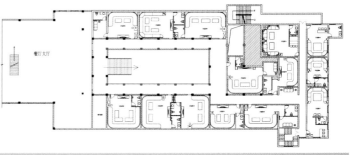

天宫内性格气质不同的包房,以奥妙般盛宴款待着深夜寂寞的回归者。色彩、灯光、线条、平面、形状、材质、节奏、音调——多元化的元素虚实辉映,幻化为不断舞动的生物灵态。天宫即幻界,这里没有墙壁、没有天花,只有无限延展的空间,容纳着无数陌生而熟悉的魂魄。

值得一提的是,宽大、舒适而有支撑力的沙发,柔软而又有厚度的地毯,成就了遨游天宫的美妙。身体的每一部分都被彻彻底底、踏踏实实地放在了那里,心灵和欲望才得以没有任何顾虑地冲出躯体,在没有边际、没有方向的天际游荡,以探索无穷的可能性。

In the Heaven Palace, the compartments of different characteristics and qualities entertain, with secret grand banquet, the lonely returners in late night. Color, lamplight, line, plane, shape, materials, rhythm and tone, all the diversified elements are shining with reality and fantasies to transform into the living and spiritual beings who are continuously dancing. The Heaven Palace is the world of fantasy, where there are no walls and no ceilings but there is the infinite and outstretched space to contain numerous strange and familiar souls.

What is to be mentioned is that the spacious and comfortable sofas with a bearing and the soft and thick carpet have produced the wonderful roaming to the Heaven Palace. Since every part of the body is put there thoroughly and completely, the spirit and desire could rush out of the body with no hesitation at all to wander at the horizon with no boundary and no direction to explore the possibility of infinity.

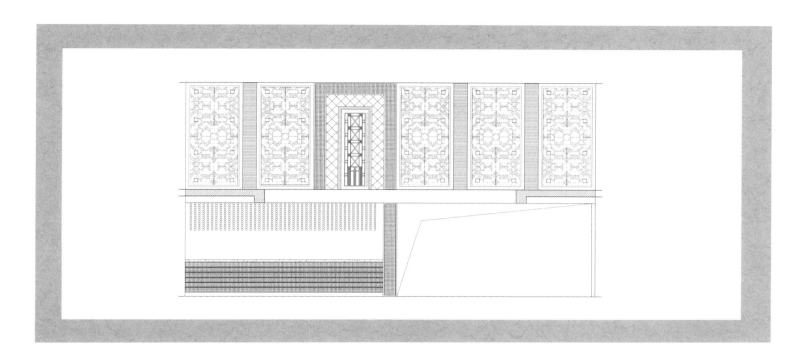

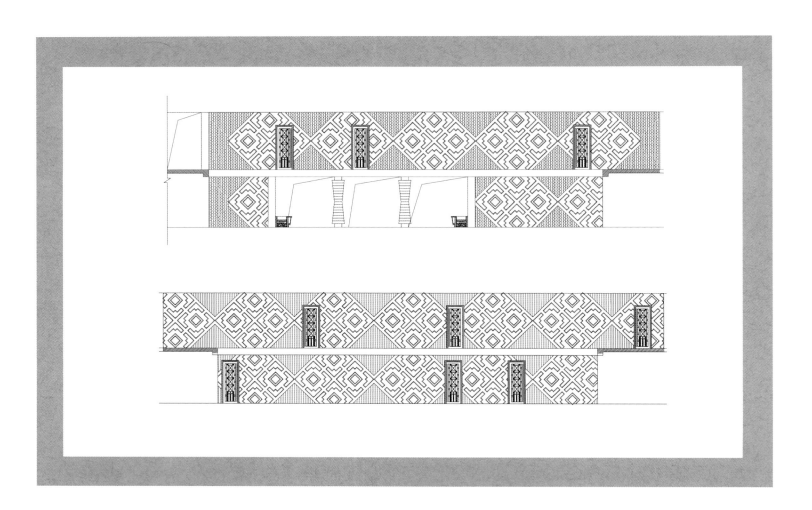

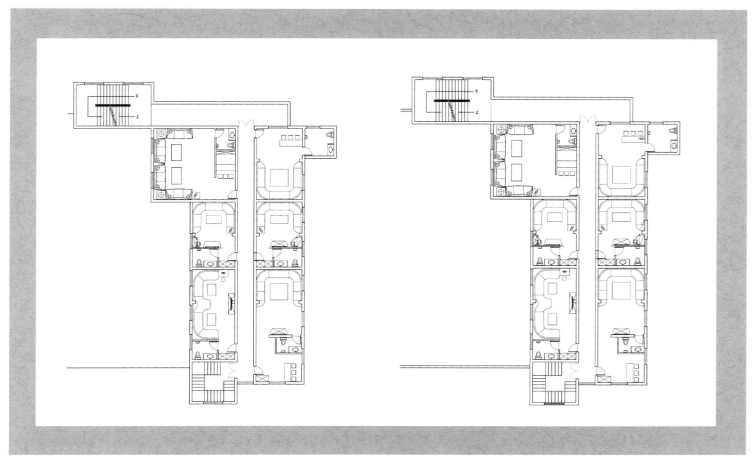

梦会所
Dream Club

项目名称 _ **梦会所**
项目类别 _ **会所**
设计面积 _ **20 000 m²**
项目地点 _ **江苏苏州**

Project _ **Dream Club**
Project Category_ **Club**
Design Area_ **20,000 m²**
Project Location_ **Suzhou, Jiangsu Province**

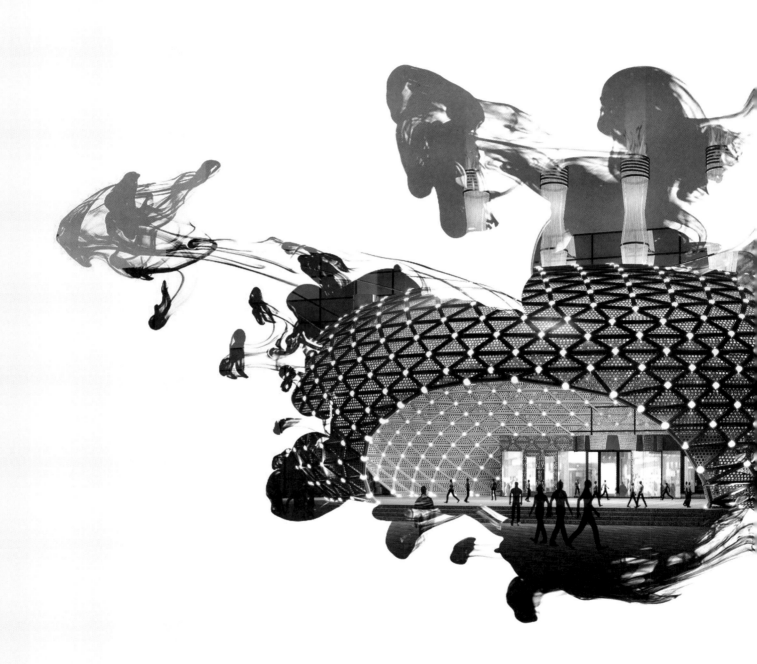

"梦"会所,位于苏州相城的湖心小岛。九栋现代建筑依水而建,水光潋滟中幻化出这个产生梦、寻找梦、体验梦、实现梦的地方。心随梦回,在这里,大家以梦沟通。

灵感来源于设计师"梦中的大自然"。他将大自然中可见的各种性灵抽象并升华,以色彩、灯光、线条、造型等瑰丽元素,展现了一个既光怪陆离又似曾相识的世界。

在这里,人以至纯至真的方式"梦归自然"。设计师突破传统的设计手法,充分应用立体投影系统、3D模拟空间技术、动漫科技、全息影像等技术,使人们成为水中的鱼、空中的鸟,自然、自由、自在地与天地间各种生灵平等存在、灵性相通。

万物自由,怡然自乐……亦幻亦真,这是一个可信、可见、可触的梦境。

梦境之美,无以伦比。

The "Dream" Club is located on the islet in the middle of a lake at Xiangcheng, Suzhou. Nine blocks of modern buildings are constructed near the lake. In the glisten of the rippling lake water, it is transformed into a place to create, seek, experience and realize dreams. Minds return with dreams. The people here could communicate with dreams. The inspiration was derived from the designer's "Nature in Dream". He abstracted and sublimated various spirits available in nature, and demonstrated, with such magnificent elements as color, light, line and shape, such a world that is weird but seems to be familiar.

The people here could "return to nature in dream" in the purest and truest way. The designer broke through the traditional design technique, and made full use of such techniques as 3D projection system, 3D simulation space technology, animation technology and holographic imaging technology to transform the human beings into fishes in the water and birds in the air, in which they co-exist equally and communicate spiritually with various living creatures in Heaven and on Earth in a natural, free and unrestrained way.

All things could be free, contented and happy... Being either realistic or fantastic, it would be a dream that is credible, visible and touchable.

The dream world is wonderful, and it is incomparable.

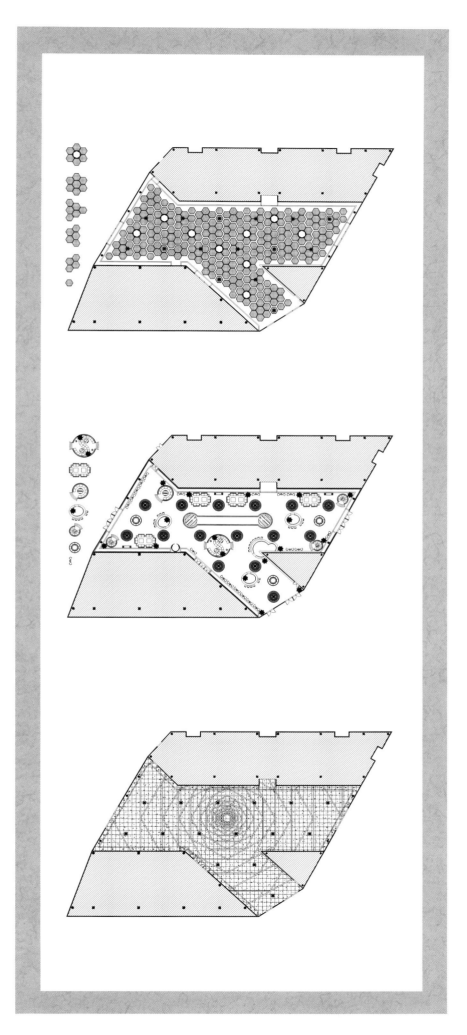

萨尔斯堡别墅
Salzburg Villa

项目名称 _ **萨尔斯堡别墅**
项目类别 _ **会所**
设计面积 _ **10 000 m²**
项目地点 _ **四川成都**

Project_ **Salzburg Villa**
Project Category_ **Club**
Design Area_ **10,000 m²**
Project Location_ **Chengdu, Sichuan Province**

古城成都,富庶繁华,山水秀丽,自古乃以"宜居"著称。民风闲适,万物悠游,总能吸引无数羁旅纷至沓来而不知返也。成都人自居蜀地,质朴地守候着经年不变的古城温柔,怡然自在地享受山水和文化带给他们的闲情逸致,不奢侈不虚夸,虔诚而又安详。当萨尔斯堡优雅的格调随着音乐之声漫至此地,与这里的气质竟可以如此契合,实在是因了这份自然与人文共融而产生的宁静悠然。

The ancient city of Chengdu is rich, prosperous and beautiful, and it is famous since ancient times for its "livability". Since its folk customs are comfortable and all things there are leisurely, it could always attract countless outside travelers to go there in large numbers and enjoy themselves so much as to forget to leave. Chengdu people feel honored to live in Sichuan. They naively keep watch on the changeless gentleness of the ancient city and happily enjoy the leisurely and carefree mood brought to them by the landscape and culture, without luxury and exaggeration but with devoutness and sedateness. When the elegant style of Salzburg roams here with the sound of music, it could so expectedly correspond to the temperament here. This is actually because of such a tranquility and leisureliness produced from the communion of nature and humanity.

萨尔茨堡会所的音乐厅，秉承了经典的欧洲教堂风格。美丽端庄的圣母，辐射出母性的圣洁光辉，人的性灵豁然间在纯净中升腾。一架黑色钢琴静静伫立在圣母圣婴图下，等待天使之手临幸，发出美妙的乐音。三面环窗，简洁流畅的拱形线条诠释了圆融和谐的理念。音乐厅中暖光熠熠，所有的一切在与音乐的共鸣中，升华并融为一体。

The concert hall of Salzburg Club upholds the classic European church style. The beautiful and elegant Holy Virgin radiates the holy radiance of maternity, and the human spirit would be suddenly sublimated in pureness. A black piano, standing there quietly under the drawing of the Holy Virgin and the Christ Child, is waiting for the angel's hand favor to produce a wonderful tone. It is embraced on three sides by windows, and the concise and fluent arch line interprets the concept of syncretism and harmony. As the warm light glistens in the concert hall, all the things are being sublimated and incorporated into one entirety in resonance with music.

欧式传统的居落，古朴的原木色与淡雅的石材裹挟自然的气息，返璞归真的娴静自此拉开帷幕。坐在萨尔茨堡客厅的沙发，小酌卡布奇诺，看着壁炉里火焰柔美舞蹈。白烛、鹿角、绿植、方圆混成的吊顶，将自然的简单与人生的哲思融而为一，仿佛置身于迷雾森林，氤氲中听生命的律动，奔跑、奔跑，直至穿越森林后喜见汩汩的溪水。溪水，流淌过数千年的生命，

The European-style traditional house and the natural flavor from the simple log color and the quietly elegant stone raise the curtain since then to show the demureness of recovery of the original simplicity. Sitting in the sofa in the saloon of Salzburg and drinking cappuccino coffee, you could see the flame in the fireplace dancing gracefully. The suspended ceiling mixed with candlelight, antlers, green plant and circumference blends both the natural simplicity and the human philosophizing into one unity, when you could feel as if you were staying in the foggy forest to hear the rhythm of life in the dense fog and run with no rest until you pass through the forest and delightedly see the gurgling stream. The stream, flowing with a life of thousands of years and mixed with the channeling church bell, is calling upon our souls and bodies to wash off all the cosmetics and flow towards the new purity.

混合教堂通灵的钟声，召唤着灵魂和肉体，洗尽铅华，一起流向新的纯净。

建筑如凝固的音乐，同时凝固于此的还有时间和品味。别墅的其他设计细节，也贯穿着设计师始终不变的诉求：与自然契合，重燃一股生命的热忱；与上天通融，重获一颗纯净善良的心。经历有我和无我、有形和无形，此番荡涤、此种膜拜，便可不苟于人世许多纷争，独享自我的一隅安宁。

安然境界，亦如萨尔茨堡，亦如成都。

Architecture is something like the solidification of music. Solidified herewith are also the time and taste. The other details of the villa design are also penetrated with the designer's constant appeal: To be integrated with nature to reignite the enthusiasm of life; and to be harmonized with heaven to regain the heart of purity and kindness. If you could experience selfness and selflessness, tangibility and intangibility, such a course of cleaning and such a kind of worship, you would discard many human disputes and enjoy yourself with the tranquility.

The realm of tranquility is either like Salzburg or like Chengdu.

鸟巢文化主题餐厅

The Bird's Nest
Cultural Theme Restaurant

项目名称 _ 鸟巢文化主题餐厅	Project Name_ **The Bird's Nest Cultural Theme Restaurant**
项目类别 _ **餐厅**	Project Category_ **Club**
设计面积 _ **4 000 m²**	Design Area_ **4,000 m²**
项目地点 _ **北京**	Project Location_**Beijing**

鸟巢，21世纪全球最伟大的建筑之一，承载了奥运史、建筑史以及中华民族发展史上无数个第一。它见证了2008年那个伟大的历史时刻，成就了无数梦想的辉煌绽放，演绎了无与伦比的华美乐章，更浓缩了一个伟大民族自信崛起的历程。

远观鸟巢，整体造型丰满圆润，正像一个孕育和聚合生命的巢，在网状树枝的支撑下安静而从容，令人有温暖回归的感觉。行至近前，巨型钢架结构雄浑大气，强悍刚劲，令人只有仰视和震撼。进入巢内，没有过

The Bird's Nest is one of the greatest architectures in the world in this century, which carries countless "first" in the history of Olympic Games, the history of architecture and the history of the development of the Chinese nation. It has witnessed the great historical moment in 2008, accomplished the brilliant blossom of innumerous dreams, deduced the unparalleled magnificent movement and concentrated the course of a great nation in her self-confidence and rise.

To view the Bird's Nest in a distance, you would find that its overall modeling is of fullness and perfect fusion, which is, just like a nest to nurture and polymerize life, quiet and calm in the support of reticular tree branches, so as to make

多的装饰,单纯的空间只有"中国红"和"长城灰"的简练配搭,令人专注于场内的竞技。

在后奥运时代,在这个非凡的建筑里设计一个文化主题餐厅,对设计师而言无疑是一个巨大挑战。首先,竞技场与餐厅,是两种完全不同的功能空间,对室内设计的诉求也大相径庭。置身于鸟巢体育场中的餐厅,该如何去兼顾和融合?其次,鸟巢蕴涵的东西太多了,鸟巢文化主题餐厅要表达的东西也太多了,在极简的背景下,如何去精炼,如何去升华?

people a warm feeling of returning home. To step close to it, you could see that the giant steel-frame structure is so forceful and generous, intrepid and vigorous that it could make people produce only a sense of respect and shock. To enter into the nest, you could find no overmuch adornment but only the succinct combination of "Chinese red" and "Great Wall grey" in the simple space to make people focus on the competition on site.

It is no doubt a huge challenge for the designer to design a cultural theme restaurant in such an extraordinary architecture in the post-Olympic era. Firstly, the arena and the restaurant are two completely different functional spaces, and their appeals for the interior design differ greatly. As for the restaurant located in the Bird's Nest stadium, how should we make a concurrent consideration and communion? Secondly, the Bird's Nest contains too much things and what the Bird's Nest Cultural Theme Restaurant would express is also too much. Under such a minimalist background, how could we refine and sublimate it?

In such an architectural environment, the interior designer has finally, based on the Chinese traditional philosophical thought of "Yin and Yang opposing but inter-promoting each other", inherited the architectural design concept of "telling Chinese stories in international languages" and adopted a mode of expression that is different but of mutual benefit in the blended design of both interior and exterior spaces. One is the rough outline of master stroke, and the other is delicate and exquisite; one is downright and unconstrained, and the other is gentle and graceful. This is just the so-called "Yin and Yang inter-promoting, Yang inside Yin, Yin also inside Yang, both energy softened for harmony". In between motion and stillness, he follows the principle of Tai Chi (the Great ultimate) of "Yang generated from motion, extreme motion turning into stillness, Yin generated from stillness, extreme stillness turning into motion again; Motion and stillness, origin of each other". At last, the word "harmony" has solved all the problems. "Sympathy of Yin and Yang could make reincarnation of all things". In making such a harmony, the greatest "nest" in the world would polymerize even more breath and nurture even more

在这样的建筑环境下,室内设计师最终以中国传统的"阴阳对立相生"哲学思想为基础,承袭了"用国际语言讲述中国故事"的建筑设计理念,并采取了不同但相互补益的表达方式进行室内外空间的融合设计。一个是大手笔粗线条,一个是精致细腻;一个直率豪放,一个柔美婉约。正所谓"阴阳相生,阴中有阳,阳中有阴,冲气以为和"。在动静之间,依循"动而生阳,动极而静,静而生阴,静极复动。一动一静,互为其根"的太极之规。最终,一个"和"字解决了所有的难题。"阴阳交感,化生万物",如此调和,世间最大的"巢"将聚合更多的气息,孕育更多的希望。

以新型现代材料GRG制作的镂空网架隔断,以更亲切的语言呼应鸟巢钢质网架的建筑结构美;在灰色的大背景下,以包容性更强、更明快、更轻松的白色和透明色为主色调,再以"中国红"点睛,巧妙地传承和提升了建筑环境的内涵;餐厅内巨大的方型钢质斜柱,在不失本色的前提下,以剔透的玻璃、晶莹的水晶、庄重喜庆的红丝线以及圆满的祥云图案装饰,营造了整个空间精致、柔美、圆融、亲切、和谐的氛围。落座于此,透过整幅的玻璃窗,可以看到整个体育场,看到大屏幕回放的奥运精彩瞬间,内心可以从容地感受那时那刻的辉煌。此时的鸟巢,可以身心融入,是属于每个普通人的。

"中国红",是中国人的魂,有时热情奔放,有时含蓄庄重,经过世代承启、沉淀、深化和扬弃,弥漫着浓厚得化不开的中国情结,象征着热忱、奋进、团结和沉稳的民族品格。这是存在每个中国人潜意识里的中国文化。鸟巢是中国人东方的智慧和现代科技的完美结合,是中国文化的凝聚与释放。把红色作为鸟巢主题文化空间内唯一的彩色,它向世界坦荡地展示了一个国家的精神气质和民族自信心。此时的鸟巢,是属于每个中国人的。

古老的活字印刷、中式花格以及和平鸽在设计师的意象中进行了超时空变异,以更轻盈、更现代、更直接的方式将人类文明和中华文化的千年演变瞬间呈现,令来自世界不同角落的人在不需要任何语言的解说下,顿时可以体味出这种悠远而美好的意蕴,沉浸在人类文明恢弘气象中,心潮澎湃,血脉迸流,这是人类的共鸣。此时品赏的不仅是美食,更是醇厚悠久的文明盛宴。此时的鸟巢,不仅仅属于中国,更是属于世界的。

hopes.

The hollowed-out net rack partition made of the new modern material of GRG is used to echo in an even more friendly language with the beauty of the architectural structure of the Bird's Nest steel net rack. Under the great grey background, the dominant hue is white and transparent color that is ever stronger in comprehensiveness, more vibrant and more comfortable, which, matched with the "Chinese red" color for key points, has cleverly inherited and promoted the connotation of the architectural environment. As for the huge square steel inclined column inside the restaurant, in the prerequisite of keeping its natural color, we have used the transparent glass, glittering and translucent crystal, solemn and jubilant red silk thread and the plump lucky clouds device in adornment, so as to build an atmosphere of delicacy, gentleness, communion, kindness and harmony in the whole space. When you are seated there, you could, through the full size of casement windows, see the entire stadium and watch the highlights of Olympic Games played back on the large screen, when you could calmly experience, in your heart, the brilliance at that moment. In the Bird's Nest right now, your body and mind could be blended together and it belongs to each of the

ordinary people.

The "Chinese red" is the soul of the Chinese people. Sometimes it is an outburst of enthusiasm, but sometimes it is implicative and solemn. After generations of inheritance, deposit, furtherance and ablation, it is permeated with the profound Chinese complex that would last forever, which is the symbol of the national character of enthusiasm, advance, unity and steadiness. This is the Chinese culture that exists in the sub-consciousness of all the Chinese people. The Bird's Nest is the perfect combination of the Oriental wisdom of the Chinese people with the modern science and technology, and the coherence and release of the Chinese culture. As the "red" color is used to be the sole color in the cultural theme space of the Bird's Nest, it has demonstrated magnanimously, to the world, the spiritual temperament and national self-confidence of our country. The Bird's Nest right now belongs to each of the Chinese people.

The ancient movable-type printing, the Chinese-style lattice and peace dove have been modified in hyperspace in the image of the designer, so as to demonstrate, in an instant, the millennium of the human civilization and the Chinese culture in an even more lightsome, more modern and more direct way. With no need of any interpretation in any language, all the people in the world from different places could immediately appreciate such a distant and beautiful rhyme and charm and then immerse themselves in the tremendous momentum of human civilization with great excitement. This is the resonance of the humanity. What they taste right now is not only the gourmet food but more also the feast of the mellow and long-standing civilization. The Bird's Nest right now would belong not only to China, but also to the whole world.

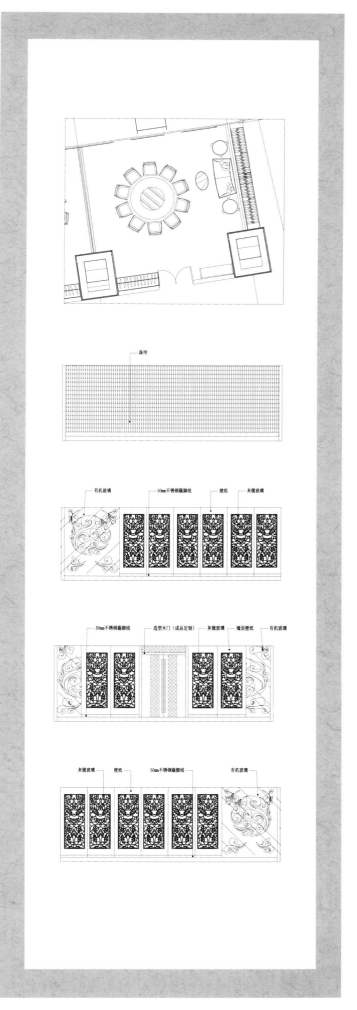

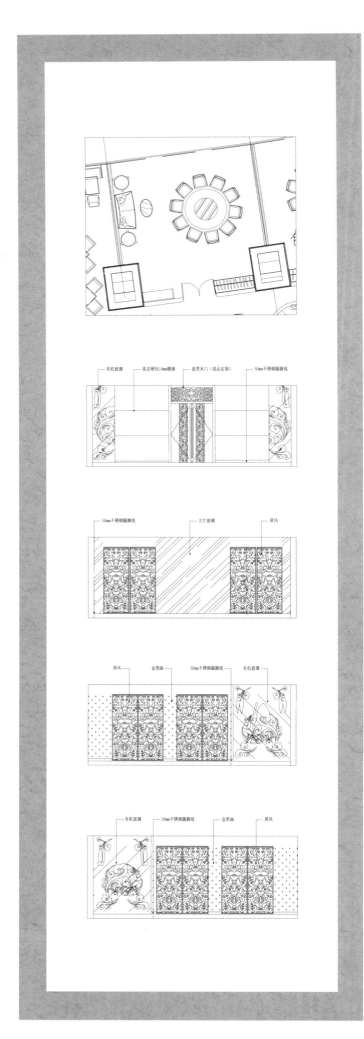

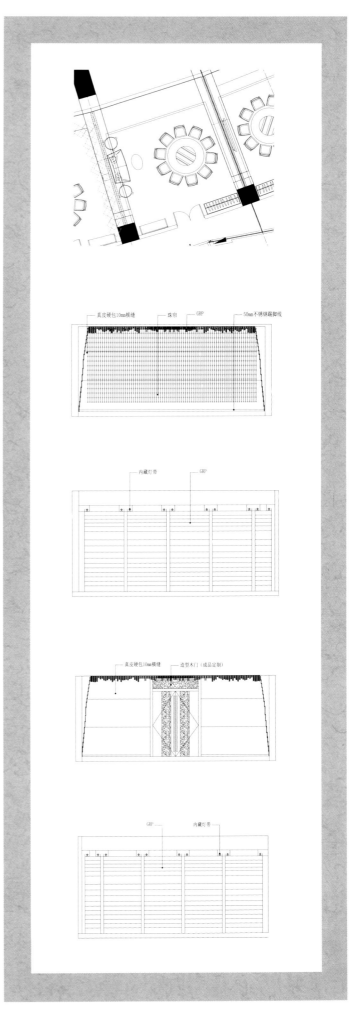

蓝湖郡别墅
Lanhujun Villa

项目名称 _ 蓝湖郡别墅
项目类别 _ 别墅
设计面积 _ 1 200 m²
项目地点 _ 重庆

Project Name_ **Lanhujun Villa**
Project Category_ **Villa**
Design Area_ **1,200 m²**
Project Location_**Chongqing**

山城重庆之于人的，不仅是一种巍峨林立之美，更在其婉约灵动的神韵。面对瑰丽风流的渝州城，本案设计师随之萌生出物我相忘、性灵交融的创作激情。于是，湖光山色之间，蓝湖郡别墅应运而生。

The attraction of the mountain city of Chongqing for people is not only the beauty of a forest of majesty, but also its graceful and smart verve. Facing the magnificent and romantic city of Yuzhou, the designer of this case has thus created the creative spark of the selflessness of nature and humanity and the communion of spirits. Hence, in the natural beauty of lakes and mountains, Lanhujun Villa was born at the right moment.

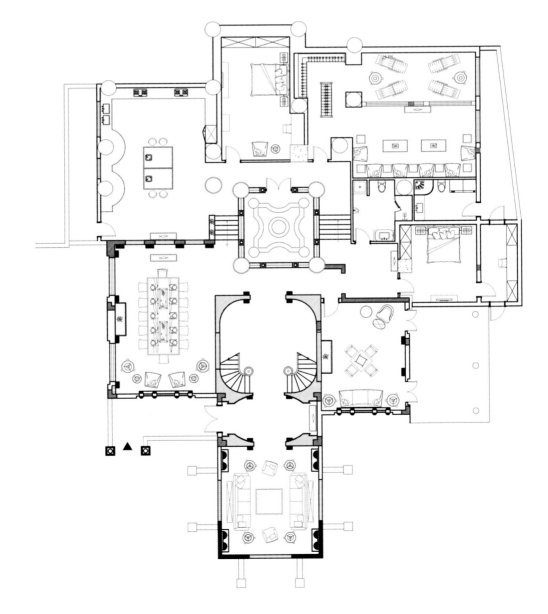

本案的设计包括建筑结构改造、园林规划、内部装饰等全部设计内容。在设计师第一次看到这个具有天生禀赋却又平凡无奇的别墅开始,便在脑中形成了大胆的设想——在湖边做一个"流水别墅"。

从"玄武探水""玉带缠腰""御水丹田"的风水格局,三面风景、临水驾风的空中卧室,到功能完善、流线顺畅的内部空间布局,清新别致的屋顶花园,再到天地贯通的女神天井,自然音容在每一个角落的渗透,以及典雅华贵的装饰风格……一个变平凡为神奇、再到传世经典的传奇,在设计师一气呵成的创作中诞生了。

The design of this case includes all such design contents as building structural transformation, landscape planning and interior decoration. At his first sight of the villa of natural endowment but simple ordinary, the designer then created a bold assumption in his mind—to make a "Falling Water Villa" at the lakeside.

From the geomantic framework of "tortoise searching water", "jade belt around waist" and "public region resisting water" and the hanging bedrooms with three sides of scenery and located near water and on the wind, to the interior spatial layout of perfect function and smooth streamline and the fresh and unique roof garden, and then to the goddess courtyard of heaven-earth coherence, you could see the natural likeness permeated in every corner and the decorative style of elegance and magnificence… A legend of transformation of ordinary to miracle and then to heritage has been produced in the coherent creation of the designer.

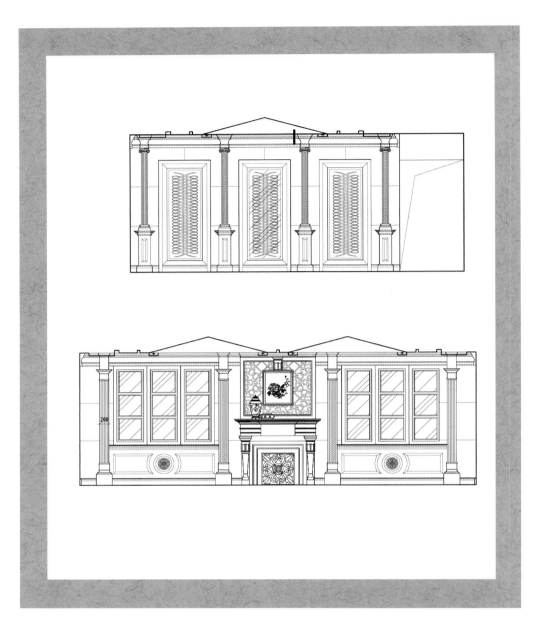

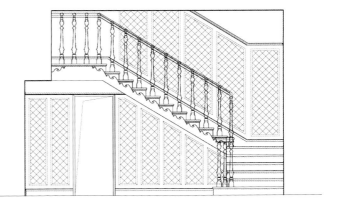
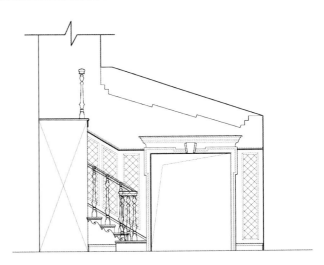

这是一个清纯脱俗、秀美灵动的有机建筑，生长在中国式田园意境与意大利托斯卡拉乡土的自然风情中，这是一个听溪语鸟鸣、睹鱼跃鸢飞、观修竹清泉的室外桃源。

设计的价值，不仅仅是从 500m² 扩展到了 1100m²，而且是一种内在品质的升华，是一种经得起时间涤荡的传承品格。

This is an organic architecture of purity, unconventionality, elegance and smartness that grows in the natural landscapes of both the Chinese-style artistic conception of countryside and the Italian Tuscan native land, and this is a land of idyllic beauty where you could hear the sound of brook and the voice of birds, to see fish leaping and birds flying and to enjoy the sight of tall bamboo and clear spring.

The value of this design is not simply only the extension from 500 square meters to 1,100 square meters, but also the sublimation of the intrinsic quality, which would be the character of inheritance that could withstand time clarification.

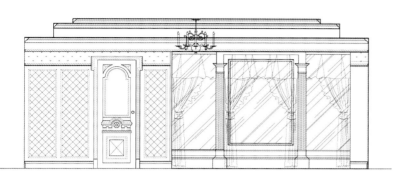
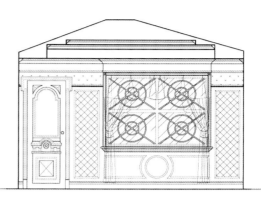

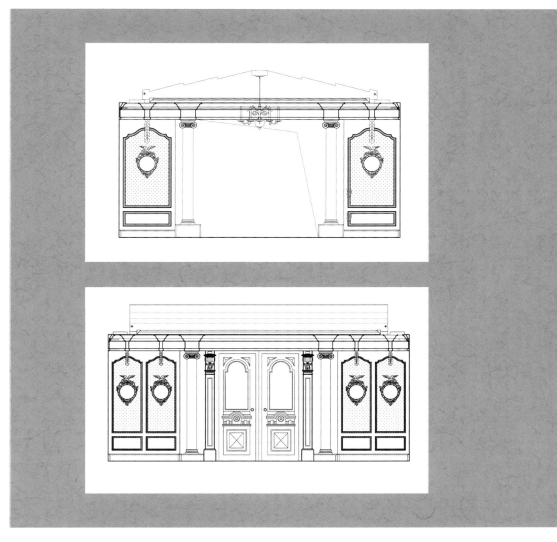

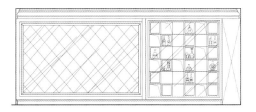

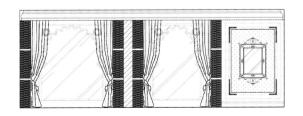
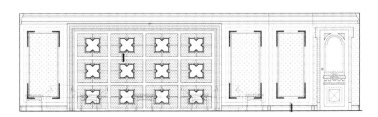

神华集团神包矿业行政办公楼及联合建筑

Office Building and United Building of Shenhua Baotou Mining Group

项目名称_ 神华集团神包矿业行政办公楼及联合建筑	Project Name_ **Office Building and United Building of Shenhua Baotou Mining Group**
项目类别_ **综合建筑**	Project Category_ **Complex Building**
设计面积_ **22 000 m²**	Design Area_ **22,000 m²**
项目地点_ **鄂尔多斯**	Project Location_**Ordos**

神华,作为世界上最大的能源企业,正如该项目所在的城市鄂尔多斯,有着雄浑宽厚的气质和海纳百川的胸怀。

能源是大自然赋予人类的厚礼,人在自然中,人也是自然。对于这份厚礼,我们应给予足够的珍视。辽阔的鄂尔多斯草原造就了本项目的宏大风格。集团大厅的菱形交叉透明玻璃屋顶,与蔚蓝的天空直接对话,这是对自然之光的渴望和珍惜。大自然赐予我们的,是丰富的能源,更是生存的希望和生命延续的勇气。阳光房荡漾着蓝色"水波"的球形屋顶,让人仿佛如鱼儿般在水下与阳光捉迷藏,惬意地享受着自然的美好,一刻也不敢懈怠生命。

浴室空间的设计是最具人性化的,试图带人们走进那片自然幻境。蓝绿色调混合清澈的水色,朵朵金色"莲花"似乎幽香扑鼻。氤氲的气息,不知是雾气还是仙气,在此洗浴,可享全然放松的快乐,更可在半仙境半人间之中重获心灵的涤荡。这是设计师在建筑的人文性方面所用的巧妙心思。

在建筑的其他设计细节方面,设计师也慷慨地运用大构图、长线条,方圆混成,自然和人文巧妙统一,物我相融,那雄浑的草原霸气就此便活灵活现了。

Shenhua is the largest energy supplier in the world. Just as the city of Ordos where this project is located, it has a vigorous and generous temperament and a broad and compatible bosom.

Energy is the lavish gift that Nature has endowed to humanity. Human beings live in nature, and humanity is also nature. As for such a lavish gift, we should attach great importance to it. The vast Ordos prairie contributes to the grand style of this project. In the lobby of the Group, the diamond cross transparent glass roof could have a direct dialogue with the blue sky. This is the aspiration for and appreciation of the natural light. What Nature has endowed to us is the continuous energy, and, furthermore, our hope for existence and our courage of life continuation. The rippling blue "water waves" on the spherical roof of the sunlight house could make people as if they were fish playing a hide-and-seek game in water with the sunshine to leisurely enjoy the beauty of nature and dare not to slacken life.

The design of the bathroom space is of the most humanity, which tries to bring people into that natural dreamland. The blue and green colors are mixed up with the limpid water color, and the golden "lotus flowers" seem to perfume a delicate fragrance. The dense breath might be either mist or fairy air. If you take a bath there, you could enjoy yourself with complete relaxation and happiness and regain the spiritual cleanness in such a situation of half-fairyland and half-humanity. This is the clever mind of the designer in respect of the architectural humanity.

In respect to other details of the architectural design, the designer has also generously used the grand composition and long lines to make the integration of square and circle and the clever unification of nature and humanity. With such a communion, the vigorous domination of the grassland could thus be demonstrated in a vivid way.

Sales Center of Beauty Crown Seven-star Hotel

美丽之冠七星级酒店售楼中心

项目名称 _ 美丽之冠七星级酒店售楼中心	Project Name_ Sales Center of Beauty Crown Seven-star Hotel
项目类别 _ 会所	Project Category_ Club
设计面积 _ 1 500 m²	Design Area_ 1,500 m²
项目地点 _ 海南三亚	Project Location_ Hainan, Sanya province

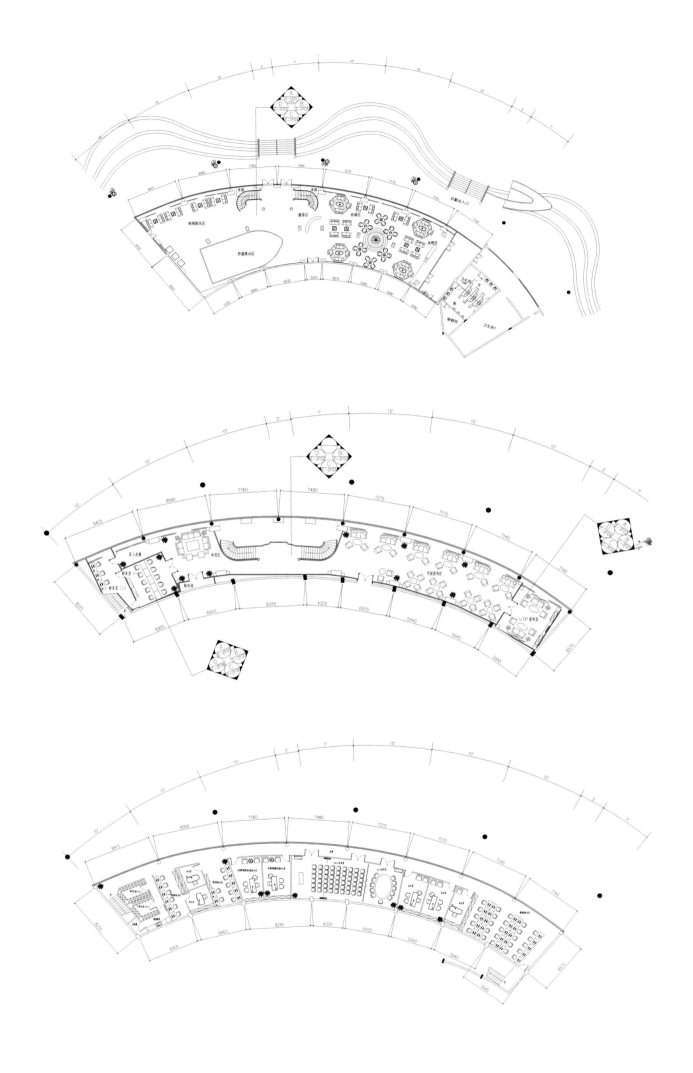

依山傍水、金色霞光中，一座巨型乳白色膜盖，犹如一朵巨型花冠闪现于万绿丛中，伴着徐徐微风，于海天一色中摇曳绽放，它就是三亚"美丽之冠"。

"美丽之冠"是世界上第一座以美丽为主题的王冠型建筑。之所以称其为"冠"，不仅仅是来自于世界各地的美丽女人汇集在这里并展示美，而且因为它是一种至高境界的美，是在天地大美之中，融合人性之美、文化之美的齐一醇和之美。本案就是在这样一个美丽神话中诞生的。

Near the mountain and by the water, and in the golden sunglow, there is a giant ivory membrane cover, which is, like a giant corolla flashing on the backdrop of greenery, swaying with gentle breeze to blossom in the sea melted into the sky. This is just the "Beauty Crown" in Sanya.
The "Beauty Crown" is the first, in the world, crown-shaped architecture themed with beauty. It is called "Crown" not only because the beautiful girls from all over the world have gathered there to show their beauty, but because it should be a beauty of supreme realm, and an integrated and harmonious beauty to blend the beauty of humanity and the beauty of culture in the grand beauty of heaven and earth. This case is just born in such a beautiful myth.

新古典欧式风格,融汇时尚与经典,将异彩纷呈的"美"演绎得淋漓尽致。雕花藻井的雍容华贵、卷花楼梯的华贵飘逸、莎安娜壁炉的端庄典雅、罗马柱的亭亭玉立、做旧金箔的低调奢华、茶钻马赛克的晶莹剔透、水晶吊灯的纯净灵动……

走进美丽之冠,你会惊叹,不知是世界小姐赋予了她美丽的精髓,还是自然之美的气息透彻熏陶,秀外慧中的她美得坦荡,美得率真,美得和谐,美得永恒。

The new classic European style, blending fashion and classics into one, has deduced the colorful and extraordinary "beauty" incisively and vividly. The elegance and luxury of the carved caisson, the nobleness and grace of the scroll stairs, the modesty and elegance of Royal Botticino fireplace, the slimness and grace of the Roman column, the low-key and luxury of the gilding done-old, the crystal clearness of tea-colored diamond mosaic, the purity and smartness of the crystal droplight…

When you enter into the Beauty Crown, you would marvel that, either because Miss World has endowed her with beautiful essence or because she has been thoroughly nurtured by the smell of natural beauty, the pretty and intelligent girl is really a beauty of magnanimousness, sincerity, harmony and eternity.

后记 Postscript

熟悉张鸿的人都知道其人好味，最喜欢大雾天约上三朋四友在黄角树下吃重庆老火锅。麻辣的醇香弥漫在浓重的雾气中，偶尔听到邻桌断断续续的猜拳行令声，循声望去却似有似无。情不自禁打个嗝，整个人都"通了"、"空了"、"松了"，"化了"，此刻的他就像那棵从土里长出的黄角树，不断地向上伸展，没有边际……

这就是他最超脱的放松方式。做设计时他总是灵感迸发、激情满满，好像永远不觉得累，是个十足的"设计狂人"。但是设计之外的路走起来却太艰辛，苦辣酸甜百味俱足，很多人都退了，他却是个另类，用他自己的土话说就是"抵拢不倒拐"，也许是"树子还没有长到头儿"，他没有理由改变方向。只是，在累了、倦了或者受伤后，他会避开世俗的侵蚀和诱惑，回归到这麻辣雾气中做一棵树，暂时地麻痹和休养，排空一切杂芜，汲取土地的力量，然后更加努力地去生长。

太多的事情要做了。现在正是民族文化复兴的时代，他总感觉自己有一种社会责任感，太多的项目因不负责任的设计而流俗，太多的学生随波逐流不知所依……面对这些，他别无选择，注定要坚持。

这本专集所收录的，是他多年来积累的设计方案，其中虽然有些因设计之外的问题遭遇阻隔未得实施，但通过它们，我们可以进一步理解设计师最基本的责任心是认真对待每一个设计方案，而方案承载了设计的灵魂，这些都说来简单却又弥足珍贵。同时，这些作品中还蕴涵着一种不断向上生长的状态，一种淡泊而执著的精神，一种"也无风雨也无晴"的境界，就像那棵黄角树，当然，还有一种难耐却又必须耐得住的寂寞。

底蕴沉香，香飘万里。他相信，纵使浓雾迷离，总有"爱味者"会寻香而至。

张亚莉 Zhang Yali

运营总监
清华大学工商管理硕士
高级项目管理师
十余年公司及大型项目管理运营经验

Operation Director
Master of Business Administration (MBA) of Tsinghua University
Senior PMP
Over ten years' experience of management operation for companies and large-scale projects

Those people familiar with Zhang Hong all know that he is a man who has a fancy for taste. He prefers to invite some friends to Chongqing old chafing dish under a yellow horn tree in foggy weather. The dense fog is penetrated with the hot and spicy mellowness. Occasionally, he may hear the intermittent voice of the finger-guessing game at the nearby table, but it might become vague and indistinct when he follows the sound. With an unconscious hiccup, his whole body would become "open", "empty", "relaxed" and "melted", at this moment, he feels like stretching upwards with no interruption and no boundary just as the yellow horn tree growing out from the soil.

This is his most unconventional way for relaxation. In doing design, his inspiration always bursts out with full passions, as if he has never felt tired. He is a pure "design maniac". But his road outside design has been too hard, which is full of different tastes and flavors of life. Many people have retreated, but he is a maverick. As described by himself in his local dialect, he would "never turn around until arrival". Perhaps because "the tree has not yet grown up", he has no reason to change his direction. However, when he feels tired or injured, he would avoid the secular erosion and temptations and return to the hot and spicy fog to become a tree, in which he could make a temporary paralysis and recuperation, empty all the disorders and absorb the strength of land, so as to grow himself with even greater efforts.

There are too many things to be done. In the current era of national culture renaissance, he always feels that he would have a social responsibility in respect to such a situation in which there are too many projects that have become too vulgar because of the irresponsible design and many students go with the stream but know nothing about their own directions… In face of all these, he has no choice but to persist with no hesitation.

What has been collected in this Collection is his design schemes accumulated for many years. Although some of them have not yet been put into implementation because of other obstruction outside design, we could, through them, further understand that the most fundamental responsibility of a designer should be the serious treatment of each of the design schemes. Such schemes carry the soul of the designer. All of these could be much easier said yet would be particularly valuable. Meanwhile, these works also contain the state of constant upward growth, the spirit of simplicity and persistence and the realm of "neither wind and rain nor shine", just like that yellow horn tree. What's more, there is also a kind of loneliness that is hard to tolerate but he has to.

The heavy fragrance in implication could spread thousands of miles. He believes that even if the dense fog is blurred, there must be some "taste lovers" who would come for the fragrance.